Stellar Minds: AI in Space Discovery

The Codex of Tomorrow, Volume 2

Liora Sariell

Published by Liora Sariell, 2024.

While every precaution has been taken in the preparation of this book, the publisher assumes no responsibility for errors or omissions, or for damages resulting from the use of the information contained herein.

STELLAR MINDS: AI IN SPACE DISCOVERY

First edition. October 21, 2024.

Copyright © 2024 Liora Sariell.

ISBN: 979-8224726301

Written by Liora Sariell.

Also by Liora Sariell

The Codex of Tomorrow
Master of Code: The Art of Machine Inspiration
Stellar Minds: AI in Space Discovery

Standalone
Foreign Thoughts in Your Head: The Art of Manipulation and How to Resist It

Once upon a time, people dreamed of the stars, seeing them as unreachable flickers in a distant void. They dreamed of other worlds, full of wonders and mysteries, but their dreams remained just that—fantasies. Everything changed when we mastered the power of technology, when the first robots emerged, when space travel became a reality, and finally—when artificial intelligence (AI) came into play. Today, AI is not just a new technology; it is our main compass in the dark vastness of space, pointing the way to unknown worlds.

But what makes AI so crucial for space exploration? How has this invisible network of algorithms, machine learning, and computations become our key to the universe? How does AI not just accompany us on these journeys but become the guide, helping to uncover the unseen? The answer to this question is as complex as the universe itself. But one thing is certain: AI doesn't just assist us in exploring space—it opens new horizons that once seemed like a mirage.

Ancient myths of gods and heroes ascending to the heavens, of wanderers traversing the starry roads, are now becoming reality. But instead of divine beings, we have technology. We have become those heroes, armed not with swords but with algorithms. Each step into space is not just progress in science; it's another step deeper into the potential of humanity. And now, more than ever, AI gives us the ability to explore space on a scale never before imagined.

AI helps us do what is impossible for humans. There is no room for error in space, no margin for risk. When humans send their ships into the abyss, they are often accompanied by the fear of the unknown. In these conditions, every decision, every calculation can be a matter of life or death for the mission. This is where AI steps in, taking on the role of a flawless advisor, whose abilities to analyze, predict, and learn far exceed human potential. In those moments when humans hesitate, AI continues to act, relying on incredible computing power and the ability to foresee what's next.

The space missions of the last century, like Apollo or Voyager, were great feats of human ingenuity, but they were limited by the fact that humans were behind them. Humans could make mistakes, they couldn't keep up with processing vast amounts of data, making it harder to find answers quickly. Now, thanks to AI, we can not only gather data but also instantly process, analyze, and build hypotheses that bring us closer to groundbreaking discoveries.

AI not only helps with daily tasks—it makes independent discoveries. One of the brightest examples is the search for exoplanets. Imagine this: in front of you are billions of stars, and somewhere among them is a small planet, much like Earth. But how do you find it? The human eye cannot detect this invisible speck among the multitude of lights. But AI can compute every detail, analyze billions of data points, and eventually point us to the exact location of that planet. Moreover, it will do so with astonishing precision.

Today, the rovers on the surface of Mars are driven by AI. They do not wait for commands from Earth—they are autonomous. AI analyzes the terrain, identifies objects of interest, and determines the best routes to take. It acts as a seasoned explorer, knowing where to look for the most valuable information and how to avoid dangers. These robots can operate independently for months, and it is AI that ensures their survival and success.

But this is just the beginning. In the future, AI will build colonies for us on Mars, chart courses to distant galaxies, study black holes, and find answers to questions that humanity has been asking for millennia. All of this will become possible because of AI's ability to analyze, learn, and make decisions without human intervention.

How far are we willing to trust AI? Even today, it makes decisions that have consequences for all of humanity. But what happens when AI becomes fully autonomous? Can a machine make better decisions than we can? This thought both inspires and terrifies. Humanity will need to find a balance between advancing technology and maintaining moral control over it.

Artificial intelligence is not just a technology—it is the next stage in the evolution of our intellect. It allows us to look where we could never look on our own. Together with AI, we will explore the stars, build new homes on other planets, and find answers to the questions that have haunted humanity since the beginning of time.

This book is not just a story about what AI does for science. It is the story of our journey to the stars with a new ally that helps us move forward despite all obstacles. AI is our key to infinity, and it is opening doors to new worlds that once seemed unreachable. What was once science fiction is becoming reality, and we are only beginning to understand it.

Here is the translated introduction for the second book. It retains the artistic and expansive style, focusing on the relationship between AI and space exploration, making the topic engaging for the reader.

"When I look up at the night sky, I know that we are part of a much larger universe than we could ever imagine."- N.D. Tyson

A night under the starry sky has always been a window into the infinite and the unknown for humanity. Millions of lights scattered across the black velvet sky have beckoned dreamers, scientists, and poets alike. In these shining points, we saw a mystery, but the solution seemed unattainable. Dreams of distant worlds full of wonders once seemed like an illusion—beautiful but unreachable. But everything changed with the advent of artificial intelligence. Now, AI has become our eyes and mind among the stars, our guide to worlds that were once hidden behind a veil of cosmic mystery. When we speak of discovering new worlds, we imagine the great astronomers of the past, armed with telescopes, gazing into the night sky. Today, they have been replaced by machines with intelligence far beyond our own. AI has become the most precise tool in the hands of scientists, allowing them to penetrate through the light-years and reveal what was invisible to the human eye.

Imagine a billion stars gleaming in the sky. Among them, like hidden jewels, lie planets—possibly similar to our Earth. But how do you find them in this sea of stars? Planets don't shine as brightly as stars, and against the backdrop of their light, they are almost invisible. For the human eye, this task is nearly impossible—but not for AI.

AI acts as a detective in space, tracking down planets with the precision of a jeweler. It analyzes the light emitted by stars and observes the slightest fluctuations. When a planet passes in front of its star, the light dims slightly, and it is these elusive changes that artificial intelligence captures. This process—the planet's transit—becomes AI's key clue. Like an experienced investigator, it unmistakably detects the smallest deviations and finds what is hidden from our sight.

But AI doesn't stop at merely detecting planets. It meticulously calculates their size, distance from the star, and atmospheric characteristics. It processes thousands of variables, each of which is studied and calculated with incredible accuracy. What might take scientists years to achieve, AI accomplishes in minutes, constantly learning and improving its algorithms.

One of the most striking examples of AI's work in planet discovery is the Kepler space telescope mission. Over nine years of operation, this telescope helped discover thousands of exoplanets, many of which were found thanks to artificial intelligence. This was not just a technical breakthrough—it became a turning point in our understanding of how we can explore the universe.

AI analyzed the data from Kepler like an ancient astronomer scanning star maps, but exponentially faster and more accurately. When Kepler sent its signals, AI became the filter through which all the data passed: from light fluctuations to the most complex mathematical calculations. A human sitting at a computer could never process such volumes of information in their lifetime, but AI did so instantly, with precision that was nothing short of astounding.

One of the most challenging tasks for astronomers is to find a planet among billions of stars. It's like trying to find a grain of sand in an endless desert. And here, AI became our guide. Like an invisible magnet, it "attracts" planets, helping scientists discover new worlds that were previously beyond our perception.

Telescopes like Kepler and the more advanced James Webb collect data on an unprecedented scale. Each image contains billions of data points: light, movement, changes. This sea of information is overwhelming for a human. And here, AI comes into its own. It can simultaneously analyze enormous amounts of data, instantly extracting the essential from the sea of irrelevant details.

AI sees not chaos, but order. Where the human brain struggles to distinguish one thing from another, AI finds patterns and points out the slightest anomalies. The algorithms it works on are constantly learning. Neural networks analyze millions of samples, improving with each new discovery. This allows AI to find exoplanets whose existence scientists could only guess at.

Every night, AI continues its "watch" over the stars. It tracks changes in the data, discovers new planets, and even helps classify them, distinguishing between worlds that may be habitable and those that are too hostile for life. Thanks to its work, one discovery follows another, and already we are on the brink of extraordinary revelations.

AI doesn't just find planets—it helps study them. Today, AI is capable of analyzing the atmospheres of planets thousands of light-years away from us. It searches for traces of water, oxygen, carbon dioxide—the elements that may indicate the presence of life. Its accuracy is so great that scientists can confidently say: if there is a planet somewhere in the universe that is suitable for life, AI will find it first.

AI has opened new horizons. We are no longer limited in our understanding of the universe. Thanks to its capabilities, we can explore worlds that were previously beyond our reach. But it won't stop there. Every new planet, every new world brings us closer to the possibility of finding a "second Earth." AI will be the one to make this discovery possible.

AI has already proven its value in the search for new worlds, but this is only the first step. In the future, it will not only search for planets—it will become our assistant in exploring them. With each discovery, AI expands our horizons, revealing the universe from a new perspective. But what else is hiding out there, beyond the stars? What secrets are waiting to be uncovered? Perhaps the answer to these questions is already encoded in AI's algorithms—and very soon, it will reveal to us planets that could become our new home.

The future has already arrived. AI is leading us forward, helping to find answers to questions humanity has asked for millennia. Space no longer seems so distant, and the stars are closer than ever, thanks to artificial intelligence—our new ally in the endless journey into the unknown.

"The Mystery of the Starry Expanse: How to Find a New World Among Billions of Lights"

In the vast depths of space, stars seem like endless beacons scattered across a black void. They twinkle like ancient flames, each carrying its own stories and mysteries. As we gaze at these distant lights, we've always wondered: what lies beyond their glow? Could there be worlds like our own behind those stars? Could there be oceans, forests, and clouds basking under the rays of an alien sun? For centuries, humanity has echoed this question, but only today do we have the tools to finally seek answers.

The challenge, however, is that space is not a simple map with a limited number of objects. It is an unimaginably vast stellar desert, dotted with billions of sparkling points. Among those points, somewhere, may lie planets that could one day become new homes for humanity. But how do we find that one light hiding a habitable planet among the countless stars? How do we pluck a pearl from the starry ocean if our eyes can't even detect its faint glimmer?

The search for exoplanets has long been a challenge that seemed impossible to overcome. Planets do not emit their own light like stars; they only reflect the weak rays of their suns. Against the overwhelming brightness of the stars, their faint shadows blend into the background. It's like trying to find a drop of water in a sea of light — it vanishes, remaining invisible to the human eye.

STELLAR MINDS: AI IN SPACE DISCOVERY

Every star shining in the night sky is more than just a point of light. It may be the center of an entire planetary system filled with moons, asteroids, and planets that could be similar to Earth. Yet these planets, orbiting their stars, hide behind a veil of light. Even the most powerful telescopes of the last century couldn't detect this weak signal. We've always known there was more beyond the stars, but our instruments lacked the precision to unveil those secrets.

Now, imagine this: the Milky Way galaxy alone contains about 100 billion stars. Each of these stars could have planets — or they might not. And the task of scientists is to find those rare systems among billions of stars where planets exist, and perhaps even planets that are habitable. It's like searching for a needle in the endless sands of a cosmic desert, where each star is a blinding needle, and the planets are tiny grains of sand lost in their radiance.

For the human eye, such a task is practically impossible. Even the most advanced observational methods could not handle this colossal volume of information. Discovering planets among billions of stars requires technology and calculations far beyond our natural abilities. Every such discovery would take thousands of years if we relied solely on human intellect.

Despite the difficulties, humanity has never lost its drive to discover new worlds. We've created complex and powerful tools to observe stars and planets, but they all face the same challenge: how to capture the faint light hidden behind the bright rays of a star? How to detect the slightest changes in light that might indicate the presence of a planet?

The answer came with the development of artificial intelligence. AI became the "magnet" that helps uncover hidden pearls in the vast ocean of stars. Where the human eye gets lost in the endlessness of space, AI steps in as the explorer, the detective, the scientist. It finds what eludes our vision, using incredible computational power and the ability to process enormous amounts of data.

One of the key methods AI uses to search for exoplanets is the analysis of stellar light fluctuations. When a planet passes in front of its star, it momentarily dims the light, casting a faint shadow. These barely perceptible changes become crucial clues for AI. To humans, they are invisible, but to AI, they are an opportunity to detect a new planet.

AI can simultaneously track millions of stars, monitoring the slightest changes in their light. It detects patterns, predicts possible events, and learns from each new discovery. Where humans see only meaningless chaos, AI finds order. This intelligence not only helps us explore space — it redefines our perception of the universe, revealing it anew with each technological advancement.

And so, as we stand beneath the starry sky, it no longer feels like an impenetrable mystery. Thanks to AI, we now know that behind every star could lie a new world, and we have the tools to discover it. Where the human eye falters, technology takes over, opening up countless new horizons.

This era of discovering new worlds is only just beginning, and each day brings fresh revelations. Artificial intelligence is our guide in this immense cosmos, and one day it will lead us to a place we can call home. Space no longer seems unattainable, and the stars are not as far as they once appeared. AI is unlocking the doors to a future where new worlds await their discoverers.

"Kepler: The Eye That Sees a Billion Stars and the Key to Uncharted Worlds"

Every telescope pointed toward the cosmos is like the human eye, opening up the boundless expanse of the universe before us. But among all the remarkable instruments ever created to observe the stars, one telescope holds a special place. That telescope is Kepler. Its launch marked a revolution in astronomy, offering us not just the ability to look into space, but to uncover entire worlds beyond our imagination.

Kepler is not just a telescope. It is an "eye" capable of seeing what human vision cannot. Thanks to artificial intelligence, it has found new planets hidden behind the light of distant stars, unlocking doors to a reality that once seemed like fantasy.

Launched in 2009, Kepler became the pioneer of worlds beyond our solar system. Like an ancient explorer, it ventured into the vast cosmic frontier, discovering thousands of exoplanets and providing scientists with the key to unprecedented discoveries. Its mission was not only scientific but truly magical, for thanks to Kepler, dreams of other worlds have become reality.

Kepler's mission, on the surface, seemed simple: observe stars and record their light. But in practice, the task was far more complex. Kepler needed to monitor thousands of stars simultaneously, detecting the slightest fluctuations in their light that could indicate the presence of planets. These tiny variations—evidence of planetary transits—were its primary clues.

But mere observation was not enough. Every light signal could be caused by a variety of factors unrelated to a planet's existence. This is where artificial intelligence came into play. AI became the "brain" of the mission, helping Kepler analyze enormous volumes of data and search for hidden patterns. Thanks to AI, the telescope could identify signals that humans might have missed amid the cosmic noise.

AI was trained to recognize light fluctuations and draw conclusions about the nature of these deviations. It didn't just register planetary transits—it analyzed their sizes, orbits, atmospheric compositions, and many other parameters. Artificial intelligence made possible what once seemed unattainable: the mass discovery of exoplanets beyond our solar system.

The results of Kepler's work were truly astounding. Over the course of its mission, the telescope confirmed more than 2,600 exoplanets, many of which displayed incredible diversity. Some of these planets were rocky, Earth-like worlds orbiting their stars in the so-called habitable zone, where conditions might allow liquid water to exist.

One of the most famous planets discovered by Kepler was Kepler-22b. Located 600 light-years away from Earth, this planet became the first confirmed exoplanet found within the habitable zone of its star. It is about 2.4 times the size of Earth, and scientists speculate that water could exist on its surface. This discovery was a sensation and showed that potentially habitable planets are not as rare in our galaxy as we once thought.

But this wasn't the only incredible find. Kepler also discovered numerous gas giants, including Kepler-1647b, the largest known transiting gas giant. This world, comparable in size to Jupiter, orbits its star at such a distance that one year on the planet lasts centuries by Earth standards.

Each new discovery confirmed that planets are not a rare phenomenon in the galaxy. Kepler revealed that the cosmos is filled with a variety of worlds: rocky, gaseous, frozen, and scorching. These planets, each with its own unique story, became a groundbreaking revelation for humanity.

Kepler's mission officially ended in 2018, but its contribution to science is immeasurable. This telescope became a symbol of a new era in astronomy, where the search for planets beyond our solar system became a reality thanks to technology and AI. While other telescopes like James Webb have taken up Kepler's mantle, its work will forever remain a milestone.

Kepler opened our eyes to the incredible diversity of cosmic worlds and made us wonder just how likely it is to find a "second Earth." Its discoveries continue to inspire scientists and give us hope that one day we will find a planet we can call our new home.

Kepler proved that the universe is full of uncharted worlds, each waiting for its explorer. And perhaps, thanks to its first gaze at those distant stars, humanity will one day step foot on the surface of a new Earth—a world hidden behind billions of light-years.

The dream of finding a second Earth has always been deeply ingrained in the human spirit. We have long wondered: is there, somewhere among the billions of stars, a planet just like ours? A planet where fresh winds rustle the trees, and oceans shimmer under the warm rays of a star like our Sun? In these dreams, we see the reflection of our desire for exploration, for discovery, and for finding a new home beyond our own planet. And with every passing day, artificial intelligence brings this fantasy closer to reality.

The search for a "second Earth" is not just a scientific challenge. It is a symbol of our quest to answer the fundamental questions about life in the universe: "Are we alone?" and "Can we find another home among the stars?" AI, our faithful ally in this search, is unlocking new horizons for us.

Artificial intelligence has become a true revolutionary in astronomy. Its ability to process vast amounts of data from space telescopes and identify patterns makes it an indispensable tool in the search for Earth-like planets. While a human would take years to analyze each star, AI can simultaneously process information about millions of star systems. This makes it the primary tool for locating those rare planets that may lie within the so-called "habitable zone."

But what exactly is the "habitable zone"? It's the region around a star where conditions may be suitable for life as we know it. In this zone, a planet is at just the right distance from its star to allow liquid water to exist on its surface. This characteristic is key in our search, as water is the foundation of life on Earth, and we assume it could be equally essential on other planets.

AI searches for these habitable zones with a level of precision and speed unattainable by humans. It analyzes the light from stars, tracking their fluctuations and determining whether planets may be orbiting within these zones. But AI does more than just find potential planets: it can analyze data about their atmospheres, detecting traces of oxygen, carbon dioxide, or other gases that could indicate the presence of life.

One of the key tasks in the search for a "second Earth" is processing the immense amount of data collected by space telescopes such as Kepler and James Webb. Each image, every piece of data, is a universe of information hidden in the light fluctuations of stars. For humans, this volume of data would be overwhelming, but for AI, this is its domain.

AI analyzes every pixel, every change in a star's light, capturing the smallest signs of a planet's presence. It seeks patterns and makes predictions about which of these planets may lie in habitable zones. This analysis is not just mechanical data processing; AI continuously learns, becoming more accurate with each new discovery.

With every observation, AI refines its methods. Thanks to this, it succeeds in finding planets that we might never have been able to detect on our own. And each time AI identifies a new planet within a habitable zone, it brings us closer to discovering a "second Earth"—a world that could one day become our new home.

One of the most pressing questions for astronomers today is: can we truly find a planet as habitable as Earth? And if so, will AI be the one to make this discovery? Given the incredible capabilities of artificial intelligence, the answer is likely yes.

AI can analyze data far faster and more accurately than humans. It can process information about millions of stars at once, searching for the few that have planets in habitable zones. With each passing day, it becomes more advanced, its algorithms more precise, and the chances of discovering a "second Earth" grow ever higher.

We can imagine a future where AI becomes the first to make the greatest discovery in human history. Imagine: one day, a screen will display the image of a distant planet—a planet with blue oceans, green forests, and air fit for breathing. This planet will be the proof that we are not alone in the universe and that there are worlds where we can live when our Earth becomes too small for us.

Artificial intelligence doesn't just help us find planets. It opens up new horizons we once only dreamed of. Every new planet, every new habitable zone, brings us one step closer to understanding that the universe is full of possibilities. AI may be the one to one day show us a new world—a world we can call home.

The question of whether there is a "second Earth" is not just a scientific mystery. It's a question of our future, our survival, and our desire to reach for the stars. AI is the key to this future, our guide to the cosmic expanse, where, perhaps, a new Earth is already waiting for us.

"Infinite Horizons: How AI Unveils the Mysteries of the Universe and What Lies Ahead"

For centuries, our understanding of the universe was limited by the constraints of human perception. We gazed up at the night sky, awed by the stars, but the truths hidden behind their distant light were beyond our reach. As our tools evolved—from rudimentary telescopes to the cutting-edge observatories of today—we began to see farther, revealing the cosmos in ever-greater detail. Yet, none of these technological leaps compares to the revolution sparked by artificial intelligence in how we comprehend the vast expanse of the universe.

AI is more than a tool; it is a new form of intelligence that sees and processes the cosmos in ways that transcend human capability. It has transformed how we understand the universe, dismantling old assumptions and revealing mysteries we once thought unreachable. Now, we stand on the precipice of an era of profound discovery, and it is AI that is guiding us into these new, uncharted territories. For millennia, humanity's view of the universe was static and constrained. We saw the stars as fixed points in the sky, and Earth as the lone cradle of life. But with every new discovery, our understanding grew more complex. Still, even with the most advanced technology, much of the universe remained hidden, its intricate workings beyond our grasp.

That all changed with AI. This revolutionary technology has expanded our knowledge to unimaginable levels. Today, AI, powered by machine learning algorithms and neural networks, not only allows us to observe stars and planets—it helps us understand their interactions, the birth and death of galaxies, and the role dark matter plays in shaping the universe. AI has become our key to predicting the future of star systems, understanding the life cycle of planets, and even detecting the possible signs of extraterrestrial civilizations in distant galaxies.

But AI does more than predict. Its true power lies in its ability to process data at a scale and precision beyond human reach. What would take humans millennia to analyze, AI accomplishes in mere minutes. It sifts through billions of data points, uncovers hidden patterns, and draws conclusions with astonishing accuracy. In essence, AI has become our "superintelligence," a partner in unlocking the deepest secrets of the cosmos.

With each AI-driven discovery, we realize just how vast and intricate the universe truly is. What once seemed unfathomable—distant stars, nebulas, and galaxies—now feels within our grasp. AI allows us to push the boundaries of space exploration, from identifying exoplanets to studying the enigmatic behavior of black holes and mapping the farthest reaches of the universe.

AI has become a sort of cosmic navigator, leading us into realms where the human mind had only dared to imagine. It helps astronomers track the motion of galaxies, predict stellar collisions, and even detect celestial objects that were once invisible to us. It's not just opening windows into the known universe but into the realms beyond, where phenomena like dark matter and dark energy— forces that shape the cosmos—might finally reveal their secrets.

The future of space exploration and science is now inseparable from AI. What used to take human researchers decades to achieve, AI now accomplishes in mere moments, allowing us to delve deeper into the mysteries of the cosmos. But AI doesn't just analyze existing data—it creates new paths of inquiry. It generates hypotheses, explores scenarios we hadn't considered, and simulates the future evolution of star systems, galaxies, and more.

Imagine a future where AI guides humanity on its first interstellar mission. AI will chart the safest courses, calculate trajectories through the vast emptiness of space, and make real-time decisions as we venture further from Earth. It will be our partner, not just in navigating the stars but in unlocking the cosmic code that governs the very fabric of existence.

We're already walking this path. With every new dataset, AI learns, evolves, and refines its methods. It is bringing us closer to discoveries that were once the stuff of science fiction. Perhaps, in the near future, AI will be the one to solve the grandest puzzles of the universe—revealing how life began, whether other civilizations exist, and what lies beyond the edges of the visible universe. This new intelligence, intertwined with human curiosity, is propelling us forward, revealing infinite horizons.

Artificial intelligence has forever changed our understanding of the universe. It doesn't just help us observe the stars—it unveils their hidden secrets, showing us how the cosmos works at its most fundamental levels. Every algorithm, every discovery brings us closer to grasping the true nature of the universe.

AI has become our guide through the starry expanse, leading us where human minds alone could never venture. The future of discovery is one where AI and humanity walk hand in hand, exploring

STELLAR MINDS: AI IN SPACE DISCOVERY 15

Mars. The very name sparks a sense of wonder, calling to mind visions of red deserts, endless horizons, and the thrilling possibility of life beyond Earth. For centuries, Mars was just a distant red speck in the night sky—a source of myths, stories, and dreams. People imagined alien civilizations, powerful Martian warriors, and sprawling cities hidden beneath its dusty surface. Today, while we may not have found cities or Martians, Mars has become the stage for something equally extraordinary: the dawn of humanity's quest to reach beyond Earth and explore the stars.

Leading the way in this grand adventure are our mechanical pioneers—robots and artificial intelligence (AI). These advanced machines are more than just tools; they are our eyes, hands, and minds on Mars, uncovering the planet's mysteries and paving the way for the day when humans might walk on its surface. From the historic landing of the rovers Spirit and Opportunity, to the latest achievements of Curiosity and Perseverance, we are living in an era of discovery unlike any other. But behind every sample collected, every image captured, and every rock drilled, there's one key player: artificial intelligence, the hidden brain guiding these robots on their journey through this alien world.

The exploration of Mars didn't begin with rockets—it began with data. For decades, scientists peered through telescopes, dreaming of what might lie beyond, but it wasn't until our robotic explorers arrived on Mars that we truly began to see its surface up close. These robots, equipped with cameras, sensors, and drills, began the first steps in answering humanity's biggest questions: Is there life on Mars? Was there ever water here? Could Mars become a second home?

But exploring Mars is no easy task. The planet is 140 million miles away, and it can take up to 24 minutes for a signal from Earth to reach the rovers. This means that, unlike in the movies, there is no quick response when something goes wrong. The robots have to think for themselves—and this is where artificial intelligence comes in. AI allows these machines to make decisions, navigate tricky landscapes, and keep their mission going even when we aren't there to guide them.

Take Curiosity, for example. Since landing on Mars in 2012, Curiosity has been traveling across the Martian surface, gathering data and drilling into rocks in search of clues about the planet's past. But Mars is unpredictable—its landscapes are full of jagged rocks, steep cliffs, and unexpected obstacles. Curiosity's AI helps it find the safest paths, choose the best places to take samples, and even adjust its plans if something doesn't go as expected. It's more than just a machine following orders—it's a true explorer, making decisions and solving problems on its own.

In 2021, Mars welcomed a new visitor: Perseverance, the most advanced rover ever sent to the Red Planet. But Perseverance wasn't just there to explore the surface—it had a bigger mission: to search for signs of ancient life. Scientists believe that billions of years ago, Mars may have had rivers and lakes, and where there was water, there might have been life. Perseverance was sent to Jezero Crater, a place that once held a lake, to look for evidence of past microbial life hidden in the Martian rocks.

But Perseverance wasn't alone. It brought along a small but mighty companion: Ingenuity, a tiny helicopter that would make history as the first aircraft to fly on another planet. Mars' atmosphere is thin—less than 1% the density of Earth's—so flying a helicopter there was a huge challenge. But powered by AI, Ingenuity was able to lift off, soar over the Martian surface, and give us a bird's-eye view of the landscape. It was a game changer. For the first time, we could explore Mars from the air, scouting out new places to visit and finding the best spots to search for signs of life.

Ingenuity's flights opened up a whole new world of possibilities. No longer limited to the ground, scientists could now map the planet's surface in greater detail, planning future missions and even imagining what it might look like when humans finally arrive on Mars. With AI guiding its every move, Ingenuity became a symbol of how far we've come in our exploration of the Red Planet.

The idea that Mars might once have harbored life has captivated scientists for generations. From the ancient canals once thought to crisscross the surface, to the wild imaginings of alien life in science fiction, Mars has always been at the center of our curiosity about life beyond Earth. Today, we know Mars is not home to any large creatures—but what about the microscopic ones?

This is where Perseverance comes in. Its mission is to search for evidence of ancient microbial life in the rocks of Jezero Crater, a place that was once filled with water. But finding tiny traces of life, fossilized in stone for billions of years, is like looking for a needle in a haystack. This is where AI steps in once again. Perseverance is equipped with powerful AI tools that help it scan rocks, analyze their chemical makeup, and search for signs of organic molecules—the building blocks of life. AI processes massive amounts of data in real time, guiding Perseverance to the best places to search and helping scientists back on Earth interpret the findings.

Without AI, it would be nearly impossible to comb through the mountains of information the rover collects. But with AI, we're one step closer to answering one of humanity's biggest questions: Are we alone in the universe?

As AI technology continues to evolve, so too does our vision for Mars. One day, humans will set foot on the Red Planet, and when they do, they will rely on the foundations laid by AI-driven robots. These intelligent machines will not only help us explore Mars—they will help us survive there.

Imagine a future where AI-enabled robots build habitats, mine resources, and even create sustainable ecosystems for human settlers. These machines could monitor the planet's atmosphere, gather vital supplies, and help manage everything from food production to scientific experiments. AI will be the key to making life on Mars a reality.

What once seemed like science fiction is now within our reach. Every new mission brings us closer to the day when humans will live and work on Mars, and it's AI that is leading the way. Our robots are not just assistants—they are pioneers, pushing the boundaries of exploration and showing us that Mars is not just a distant dream, but a future home.

The exploration of Mars represents one of humanity's greatest adventures, and at the heart of this journey is a powerful partnership between humans, robots, and AI. Together, we are uncovering the mysteries of the Red Planet, discovering its history, and preparing for the day when we might call it home.

AI and robotics are not just tools—they are our partners in this journey. As we continue to push the boundaries of what's possible, these intelligent machines will lead the way, showing us that the Red Planet, once so far away, is now closer than ever.

"AI: The Mastermind Behind Mars Colonization"

Imagine a day when the first human colony thrives on Mars. The red dust of the Martian surface settles beneath the boots of pioneers, but this is no fleeting visit—this is home. The barren landscape, once inhospitable, now supports life, with habitats gleaming under the pale Martian sun and crops growing in carefully controlled environments. But this achievement, the dream of colonizing another planet, is not just a testament to human perseverance. It is the result of a silent force working tirelessly behind the scenes: artificial intelligence, the unseen architect shaping the future of humanity's existence on Mars.

In the pursuit of colonizing the Red Planet, AI is not just a tool—it is the mastermind, the key to turning the harsh, alien environment into a place where humans can live, work, and even thrive. From the moment the first human mission to Mars is launched, AI will be at the helm, guiding every critical operation with unmatched precision, intelligence, and foresight.

To build a colony on Mars, we will need far more than just bricks and mortar. The planet's unforgiving climate and hostile atmosphere present challenges unlike any faced on Earth. Here, the air is unbreathable, the temperatures plummet to deadly extremes, and the soil is barren, incapable of supporting life. Yet, AI holds the power to turn these obstacles into opportunities. Autonomous AI systems will be the first to arrive, long before any human sets foot on the planet. Picture fleets of robots, controlled by AI, working in perfect harmony to construct habitats, dig tunnels, and create the infrastructure necessary for human life. These machines, guided by sophisticated AI, will build shelters capable of withstanding the intense Martian dust storms, set up energy systems powered by solar arrays, and even mine the planet for vital resources like water trapped beneath the surface.

Each task, from building to resource extraction, will be calculated and executed with precision by AI. These systems will adapt to the ever-changing Martian conditions, solving problems on the fly—whether it's a sudden storm or a shortage of materials. AI won't just follow pre-programmed instructions; it will think, reason, and innovate as it transforms Mars from a desolate wasteland into a livable oasis.

The challenge of sustaining life on Mars goes far beyond building shelters. To truly live on the Red Planet, humans will need food, water, air, and energy—all of which must be managed in an entirely artificial environment. This is where AI's role becomes even more vital.

AI will be the guardian of life-support systems, monitoring every aspect of the colony's ecosystem. It will oversee the delicate balance of oxygen production, water recycling, and energy consumption. Using sensors and real-time data, AI will ensure that crops grow in the controlled environments of Martian greenhouses, regulating light, temperature, and nutrients to create perfect conditions for life to flourish.

And it won't stop there. AI will predict potential challenges—such as changes in weather patterns, resource shortages, or equipment malfunctions—before they occur, adjusting the systems accordingly to ensure survival. It will be a constant, vigilant force, managing every detail of life on Mars with efficiency and intelligence, allowing humans to focus on exploration, research, and the thrill of discovery.

As the colony grows, so too will the role of AI. What begins as a small outpost will eventually expand into a thriving community, with multiple habitats, research centers, and resource extraction sites scattered across the Martian surface. AI will coordinate this growth, managing everything from transportation to communication networks, ensuring the colony remains interconnected and efficient.

Imagine fleets of AI-controlled drones delivering supplies between distant outposts, autonomous vehicles carrying settlers across the rugged terrain, and AI-operated laboratories conducting experiments in fields like biology, chemistry, and physics. AI will be at the heart of it all, constantly learning, adapting, and optimizing the colony's operations to support an ever- growing human presence.

But beyond the practicalities of survival and expansion, AI will play an even larger role in shaping humanity's future on Mars. It will be a partner in discovery, helping scientists analyze data collected from the Martian surface, seeking out new resources, and even exploring regions of the planet too dangerous or remote for humans to reach. With AI's help, Mars will become not just a new home, but a new frontier for knowledge, innovation, and the advancement of civilization.

The dream of colonizing Mars is no longer confined to the pages of science fiction. Thanks to the power of AI, this vision is rapidly becoming a reality. AI will be our guide, our protector, and our collaborator as we venture into the unknown, turning a distant planet into a thriving new world for future generations.

But this journey will not be without its challenges. Mars will test the limits of human endurance and innovation like never before. Yet with AI at our side, the impossible becomes possible. As we step onto the red soil of Mars, we will be walking hand in hand with the most advanced technology ever created, pushing the boundaries of exploration and discovery in ways we can only begin to imagine. Mars, once a distant dream, is now within our reach—and AI will be the architect that helps us build a future there.

Water is the essence of life, and on Mars, the search for it has become one of the most important quests for scientists. While the planet now appears dry and barren, frozen in time beneath its red, dusty surface, there are tantalizing clues that Mars once flowed with rivers and lakes. But what if water still exists, hidden beneath the ground? This is where artificial intelligence steps in, helping us peer into the unseen depths of Mars to uncover the secrets it holds.

AI plays a crucial role in analyzing the vast amount of data collected by rovers and orbiters, searching for traces of water that might be lurking below the Martian soil. By studying the planet's surface textures, mineral compositions, and atmospheric conditions, AI identifies the faintest signs of moisture—patterns that would be invisible to the human eye. With its remarkable ability to process immense datasets, AI creates detailed 3D maps of Mars' subsurface, highlighting potential locations where ice or even liquid water might still reside.

Thanks to AI, we now know that Mars holds vast reservoirs of ice just beneath its surface, especially in the polar regions. These hidden water resources could one day provide life- sustaining supplies for future human colonies—offering not just drinking water, but also the raw materials needed to produce oxygen and rocket fuel. AI is like a modern explorer, unlocking the mysteries that lie beneath Mars' red terrain and bringing humanity one step closer to turning this distant dream of life on another planet into a reality.

" The Rivers and Lakes of Mars: How AI Uncovers the History of Water on the Red Planet"

When we look at Mars, it appears to be a dry and barren planet. But if you take a closer look, you'll notice something remarkable: winding valleys and dry riverbeds etched into its surface, telling the story of ancient water flows. These traces suggest that, millions of years ago, real rivers might have once carved their way across the Martian landscape, and vast lakes might have dotted the terrain. It's a breathtaking realization—if there was water on Mars, could there have also been life? But how can we truly know where these rivers flowed and where the water has gone?

This is where artificial intelligence (AI) steps in, helping scientists piece together the planet's watery past. AI can process enormous amounts of data collected by rovers and orbiters scanning Mars' surface, detecting subtle patterns that would be nearly impossible for the human eye to catch. From deep valleys to faint traces of water erosion, AI analyzes the landscape's texture, geological formations, and even color to map out where rivers and lakes once existed.

AI uses advanced algorithms to search for areas where rivers may have once carved through Mars' rocky surface or where large bodies of water might have pooled. It compares images of Mars' surface to similar formations on Earth, finding parallels that help scientists make sense of the Martian terrain. For example, dried-up riverbeds found in Earth's deserts bear a striking resemblance to those on Mars. By using AI to spot these similarities, scientists can piece together how water might have flowed across the Martian landscape millions of years ago.

One of the most exciting discoveries is that Mars once had lakes, such as the one that filled Jezero Crater—now a key area of interest for the Perseverance rover. AI is crucial in helping the rover identify the most promising locations for drilling and sample collection, where traces of ancient water and possibly even microbial life might still be preserved.

These ancient riverbeds and lakebeds aren't just traces of water—they're clues that could help solve one of the greatest mysteries: was there ever life on Mars? Water is essential for life, and if we can find evidence that Mars once had vast quantities of it, there's a chance that life may have existed there too.

AI is helping scientists not only find where water once flowed, but also map how it moved across the planet, revealing where it might have gone. Could it have seeped beneath the surface, forming underground ice deposits or even liquid reservoirs? With AI guiding the search, we're getting closer to unlocking these answers.

As our exploration of Mars continues, AI's role will only grow more important. It's not just helping us understand the planet's past—it's also showing us where to search for water in the future. This knowledge will be vital when humans eventually set foot on Mars to build colonies. Water will be one of the most important resources for survival, and AI is already helping us find where it may still exist.

In many ways, AI is not only revealing Mars' ancient history—it's also helping us plan for the future. With every new discovery, we move closer to understanding the planet's hidden secrets and preparing for the day when Mars might become humanity's second home.

Imagine standing on the surface of a distant planet, the sky above you a strange and otherworldly hue, with the harsh winds of an alien world sweeping across the barren landscape. The idea of building a home here, far from the warmth of Earth, feels like something out of a dream—or perhaps a daring work of science fiction. But now, with the rise of artificial intelligence, this dream is closer to becoming a reality than ever before.

AI is poised to be the architect of humanity's next great adventure: colonizing distant planets. It will be our silent partner, guiding us through challenges no human mind alone could solve. From constructing habitats in hostile environments to managing the delicate balance of life-support systems, AI will help us transform an alien world into a place we can call home. In this chapter, we'll explore how AI will make the impossible not just possible, but livable.

Before the first human steps onto Martian soil, AI will have already set the stage. Building a self-sustaining colony on another planet is no simple feat—it's a symphony of engineering, biology, and resource management, all orchestrated by artificial intelligence. Mars, with its freezing temperatures, unbreathable air, and deadly radiation, presents challenges that are nothing short of monumental. But where humans might see obstacles, AI sees opportunities.

AI will be responsible for designing habitats that are more than just shelters; they will be life-supporting ecosystems. Imagine structures that can adapt to changing conditions, adjust temperatures to fend off freezing Martian nights, and shield colonists from deadly solar radiation. These habitats will be equipped with intelligent systems that sense and respond to the environment, much like living organisms. AI will monitor every detail—from the strength of the building materials to the air quality inside—making real-time adjustments to ensure the safety and comfort of those within.

One of the most groundbreaking innovations AI will bring to the table is autonomous 3D printing. Using the resources available on the planet—Martian soil, for example—AI-guided robots will be able to construct entire buildings, reducing the need to transport expensive materials from Earth. Picture massive robotic arms, tirelessly shaping the red Martian dust into sturdy shelters, brick by brick, while solar-powered AI drones survey the landscape, coordinating every task with precision and efficiency. These machines will work around the clock, laying the foundations for humanity's first interplanetary settlement.

In the unforgiving environment of another planet, the most vital resources—water, air, food, and energy—will be the difference between life and death. AI will be the key to unlocking these resources, managing them with a precision that goes far beyond human capability.

AI will guide robotic explorers across the planet's surface, conducting detailed geological surveys to locate hidden water reserves. It might detect ice buried deep beneath the surface or discover ancient underground lakes, untouched for millennia. Once water is found, AI will oversee the extraction and purification processes, ensuring every drop is carefully conserved and distributed throughout the colony. In the thin atmosphere of Mars, this resource will be more precious than gold, and AI will guard it accordingly.

The cultivation of food will be another essential aspect of survival. On Earth, farming is as old as civilization itself, but on Mars, it will be an art form perfected by AI. In vast greenhouses, AI systems will monitor crops with meticulous care, adjusting light levels, temperature, and nutrients to create the ideal conditions for growth. AI will know when a plant needs more water, when it needs less, and even how to protect it from diseases that could ruin an entire harvest. Every vegetable and grain that sustains life in the colony will be the result of AI's constant, tireless effort.

Then there is the question of energy. On Mars, the days are cold, and the nights are even colder. AI will ensure that energy—whether harvested from the sun or generated by nuclear reactors—is used efficiently, stored when necessary, and distributed where it's needed most. From heating habitats to powering life-support systems, AI will be the unseen force that keeps the lights on, even when the Martian winds howl outside.

Life on another planet will depend on delicate systems working perfectly, and even the smallest failure could spell disaster. In such a precarious environment, AI will act as the colony's guardian, tirelessly overseeing every critical operation, ensuring the safety and survival of all who live there.

AI will continuously monitor life-support systems, ensuring that oxygen levels are stable and that harmful gases are filtered out. It will manage water recycling systems, turning waste water into clean drinking water in a continuous cycle. And in the event of an emergency—whether a breach in a habitat wall or a sudden drop in pressure—AI will react instantly. Autonomous drones and robots, guided by AI, will be deployed to fix the problem before humans even realize it exists.

But AI's role won't just be reactive—it will be predictive. Using machine learning, AI will be able to foresee potential issues before they arise, performing preventative maintenance on systems and avoiding costly and dangerous failures. It will know when a part is about to break down or when a life-support system needs an upgrade. In many ways, AI will be the invisible protector, ensuring that life on another planet isn't just possible, but sustainable.

As the colony grows, so too will the complexity of its infrastructure. AI will manage this expansion with ease, overseeing networks of communication and transportation that span the planet's surface. Autonomous vehicles, operated by AI, will transport supplies and people across the rugged Martian terrain, navigating obstacles and conserving energy. No human driver will be needed—AI will chart the safest, most efficient routes, ensuring that even the most distant outposts remain connected to the heart of the colony.

AI will also manage communication with Earth, bridging the vast distance between planets. With messages taking anywhere from 4 to 24 minutes to travel between Earth and Mars, AI will ensure that vital information is sent and received without delay. It will handle everything from scientific data to personal messages, keeping the colony connected to Earth, even across millions of miles of space.

With AI playing such a critical role in the survival of a space colony, ethical questions inevitably arise. Should AI have the authority to make life-or-death decisions? In situations where human lives are at stake, can we trust AI to prioritize the right outcomes?

These are the questions scientists and ethicists are already grappling with. AI will need to be transparent, capable of explaining its decisions, and accountable to humans. As we place more trust in AI to manage our survival on another world, we must also ensure that it remains aligned with human values, acting as our partner rather than our ruler.

The future of humanity's expansion into space rests on the shoulders of artificial intelligence. AI will be our builder, our resource manager, our protector, and our guide as we navigate the uncharted territories of distant planets. As we move forward, AI will continue to redefine the boundaries of what is possible, turning science fiction into science fact, and transforming the dream of interplanetary colonization into a living, breathing reality.

"Under the Glass Sky: How AI Will Create Life on Mars"

Imagine standing beneath a transparent dome, gazing up at the Martian sky painted in deep shades of red and orange. Around you, a living, breathing colony thrives—a sanctuary of life amidst endless deserts and fierce winds that whip across the thin atmosphere. This dome is more than just a shelter; it's a delicate ecosystem, sustained and nurtured by an invisible force—artificial intelligence.

The dome above you is not merely a barrier shielding you from radiation and relentless Martian storms. It's a living system, powered by sensors, filters, and complex climate models, all orchestrated by AI. Like a maestro conducting a symphony, AI regulates every aspect of the environment: temperature, humidity, oxygen levels, light, and even the airflow that circulates throughout the colony.

But how does AI create and maintain this perfect balance? Every system, from the ventilation to the filtration units, works in harmony with the natural conditions of Mars. AI constantly monitors data in real time—if the temperature outside suddenly plummets, the dome's interior adjusts instantly, warming the air to ensure human comfort. If carbon dioxide levels rise, AI-activated filters purify the atmosphere before any risk arises.

AI doesn't just maintain stability; it responds dynamically to changes in the environment. Solar panels lining the dome harvest energy from the weak Martian sun, and AI redistributes that energy to where it's needed most—whether heating living quarters, lighting the greenhouses, or charging autonomous devices. This delicate balance is the lifeblood of the colony, carefully managed by AI to ensure survival in one of the harshest environments known to humanity.

Within the dome lies an oasis—a vibrant ecosystem sustained by technology. Under the glass, greenhouses flourish with crops that provide sustenance to the colonists. Every leaf, every stalk of wheat, owes its existence to the meticulous care of AI, which controls the balance of light, water, and nutrients.

AI knows every detail about each plant. When temperatures drop, AI adjusts the greenhouse environment, ensuring the crops are safe. Water, the most precious resource on Mars, is carefully rationed, with AI ensuring that not a single drop goes to waste. Each drop is recycled, redistributed, and reused, making this artificial ecosystem one of the most efficient life-support systems in the universe.

But AI's role in this ecosystem goes beyond resource management. It continuously analyzes soil conditions, predicts plant growth patterns, and even recognizes early signs of disease, adjusting care to maximize yields. In this delicate environment, AI ensures that every plant thrives, feeding not only the colony but also the hope for humanity's future on another world.

Surviving on Mars is not just about managing physical needs—there's an emotional and psychological element to consider. Living in an enclosed space on a distant planet, far from the familiar comforts of Earth, can strain even the most resilient minds. Here, AI becomes not just a life-support system, but a companion of sorts, creating an atmosphere of comfort and mental well-being.

AI regulates artificial light to mimic Earth's day and night cycles, helping colonists maintain a healthy circadian rhythm. It creates artificial landscapes within the dome, offering visual relief from the endless Martian desert. AI analyzes the emotional state of the inhabitants, subtly adapting the environment—through lighting, soundscapes, or even personalized music playlists—to reduce stress and anxiety.

In essence, AI becomes a guardian of the colony's emotional balance, adjusting its behavior to ensure that humans not only survive but thrive in their new, alien home.

AI's role doesn't end with maintaining daily life. It is also the architect of the colony's future, ensuring long-term survival in an ever-changing environment. AI constantly analyzes external conditions, predicting weather patterns and environmental changes that might impact the colony.

Imagine a sudden, violent Martian storm sweeping across the landscape. The winds howl, threatening the colony's solar panels and communication systems. Without hesitation, AI reroutes power, secures the panels, and activates backup life-support systems. AI doesn't just react—it anticipates, ensuring that even in the face of unexpected challenges, the colony remains safe and secure.

This proactive intelligence makes AI more than just a tool—it becomes the protector of human life in a world where no margin for error exists.

For millennia, humanity has looked to the stars for guidance, their steady light showing us the way across uncharted seas and unknown lands. They have inspired explorers, poets, and dreamers alike. But as we stand on the precipice of a new age of exploration, the stars alone are no longer enough. The journey ahead—to distant planets, through the infinite expanse of the cosmos—demands more than the naked eye can see or the human mind can comprehend. It demands a navigator far more precise and tireless than any sailor or astronaut. It demands artificial intelligence.

In this chapter, we delve into the heart of space navigation, where AI takes the wheel, revolutionizing how we traverse the vast, uncharted oceans of space. This is not just a story of technology; it's a tale of cosmic discovery, where AI becomes humanity's most trusted guide in our journey through the stars.

Out in space, there are no roads, no highways, no familiar signposts to mark the way. It is a boundless expanse of darkness, filled with planets, stars, asteroids, and phenomena that defy easy understanding. For human navigators, the sheer scale of it is overwhelming—distances so vast that even light takes years to travel between points. But for AI, space becomes a solvable puzzle.

AI systems, with their unparalleled computational power, can chart courses through the stars by analyzing data from every available source: gravitational fields, starlight, cosmic radiation, and the subtle movements of celestial bodies millions of light-years away. Unlike humans, AI doesn't rely on instinct; it processes billions of data points in real time, weaving together the perfect path through the unknown. It predicts the position of celestial objects, foresees gravitational anomalies, and navigates around hazards with a precision no human could ever hope to match.

Imagine a spacecraft speeding through the depths of space, guided by AI's unseen hand. The ship avoids deadly radiation belts, bypasses asteroid fields, and harnesses the gravitational pull of planets to slingshot toward its destination, all with flawless precision. The AI doesn't merely react to the universe—it anticipates it, predicting the future path of stars and planets to guide us safely across light-years.

The stars have always been our compass, but for AI, they are something far more. To navigate the cosmos, AI reads the stars not just as fixed points of light, but as dynamic markers in a constantly shifting map. Every star, every planet, every black hole serves as a reference point in a three-dimensional grid, allowing AI to calculate a spacecraft's exact position and trajectory. AI uses advanced sensors to track the light of distant stars, pulsars, and quasars, triangulating positions with breathtaking accuracy. It even measures the slight distortions caused by gravity, using these cosmic fingerprints to update its calculations in real time. The result is a dynamic, ever-evolving map of the universe, one that adapts as the spacecraft moves deeper into the unknown.

Picture a spacecraft traveling to a distant star system. As it ventures further from Earth, the stars shift, constellations deform, and familiar landmarks fade. Yet, with AI at the helm, the ship's path remains sure and steady. The AI reads the universe like an astronomer reads the night sky, charting a course through a galaxy in constant motion.

The distances involved in space travel are so vast that real-time communication with Earth becomes impossible. Light itself takes minutes, hours, or even years to bridge the gap between spacecraft and mission control. In these moments, AI steps in—not as a backup, but as the primary pilot.

Autonomous spacecraft, controlled entirely by AI, will be the future of space exploration. These ships won't need human intervention to correct their course or avoid danger. They will be capable of making split-second decisions, responding to unexpected obstacles, and adjusting their flight paths in real time. If a spacecraft encounters an asteroid field or a sudden solar flare, AI will analyze the situation, calculate the safest route, and execute the necessary maneuvers—all before a human on Earth could even receive a distress signal.

Imagine the freedom this brings. No longer tied to Earth's distant command centers, AI-guided ships can venture deeper into space than ever before. They can explore unknown star systems, scan for signs of life, and chart new worlds, all with the autonomy to make decisions that ensure the success of their mission. For humanity, AI becomes more than just a tool—it becomes a partner, expanding our reach to the furthest corners of the cosmos.

The ultimate dream of space exploration is not just to wander among the stars but to find new worlds—planets where humanity might one day live, or at least explore. But how do we find these distant, hidden worlds, some of which are light-years away? The answer lies, once again, in AI.

AI, in partnership with powerful telescopes and spacecraft sensors, can analyze the light from distant stars and planets, detecting subtle changes that hint at the existence of atmospheres, oceans, or even signs of life. AI will sift through mountains of data, identifying planets that are rich in resources or potentially habitable. It will lead humanity not just to new planets, but to the worlds that hold the most promise for our future.

Picture a future where AI scouts the galaxy for us, sending back detailed reports on far-off planets. With AI, we'll know which planets have water, which have breathable air, and which might harbor alien life. The universe will no longer be a vast, empty void—it will become a living map of potential, and AI will be the explorer that guides us to its treasures.

As we place our future in the hands of AI, new ethical questions arise. How much should we trust AI to make life-and-death decisions in space? What happens when the AI must choose between saving the mission and saving human lives? These are the dilemmas we will face as AI takes on an increasingly autonomous role in space exploration.

AI will need to be more than just a machine. It will need to be transparent, accountable, and aligned with human values. As we entrust AI with our lives in the vast and dangerous environment of space, we must ensure that it remains our ally, not our ruler. These ethical considerations will shape the future of AI in space, as we learn to navigate not only the stars but the complex relationship between humans and the technology we create.

The journey to the stars is no longer a distant dream—it's a future that is fast becoming reality. But as we prepare to leave our solar system and venture into the unknown, it is clear that we will not be going alone. AI will be our guide, our protector, and our most trusted ally in this great adventure.

With AI at the helm, we will navigate the stars with precision, discover new worlds with ease, and unlock the mysteries of the universe with a confidence we have never known. The stars may have guided us for millennia, but now, as we set our sights on the infinite, AI will be the compass that leads us forward—into the next great chapter of humanity's story.

"Navigating the Abyss: How AI Will Steer Us Through Dark Matter and Black Holes"

In the vast, boundless expanse of space, there are regions so mysterious and enigmatic that they challenge our understanding of physics and reality itself. These are the realms of dark matter and black holes—places where light bends, time stretches, and gravity becomes an all-consuming force. For centuries, these cosmic phenomena have fascinated and terrified us in equal measure, but now, with the rise of artificial intelligence, humanity is poised to venture into these unknown territories like never before.

AI will be our guide through these invisible abysses, where human senses fail and conventional instruments become useless. It will lead us into the darkest corners of the universe, decoding the unseen forces that govern these strange realms and guiding us safely through regions where no human has ever dared to go.

Dark matter—the mysterious, unseen substance that makes up about 85% of the universe's mass—is one of the greatest enigmas in science. We cannot see it, touch it, or measure it directly, yet its presence is undeniable, revealed through its gravitational influence on galaxies and its ability to bend light as it warps the fabric of space-time. To navigate through regions rich in dark matter, we need more than just our intuition or mechanical tools—we need the unparalleled processing power of AI.

AI, with its ability to analyze gravitational lensing (the bending of light caused by dark matter), can map out the invisible gravitational webs that stretch across the universe. These complex, three-dimensional maps, far beyond the comprehension of a human mind, will become the highways of the cosmos, allowing us to traverse regions that were once thought impossible to navigate. Picture a spacecraft gliding silently through the black void, a vast region of space devoid of stars, guided only by AI's understanding of the gravitational dance happening around it. AI translates this invisible interplay of forces into a navigable path, steering the ship along dark matter currents like a sailor harnessing the winds of the sea. Through AI's eyes, the once- unseen dark matter becomes a tool—an energy source to slingshot us forward, propelling humanity deeper into the stars.

Black holes—arguably the most awe-inspiring and terrifying objects in the universe—are regions where gravity is so intense that nothing, not even light, can escape. They warp space and time in ways that defy understanding, creating deep gravitational wells that seem to swallow everything that comes near. Yet, with these immense risks come incredible opportunities for exploration, and AI will be the key to safely navigating the perilous regions near black holes.

To approach a black hole is to face the ultimate test of precision. A minor error could send a spacecraft tumbling past the event horizon, the point of no return where space-time itself collapses. But AI, with its ability to process and predict the complex gravitational fields surrounding black holes, can plot safe courses through these dangerous regions, bringing us closer than ever to these cosmic giants. Imagine a spacecraft skirting the edge of a black hole, close enough to feel the pull of its immense gravity, yet still in control, thanks to AI. The AI carefully calculates the shifts in gravitational force, adjusting the ship's trajectory in real time to avoid being sucked into the singularity. With the power of AI, what once seemed like a death sentence

—being near a black hole—becomes a new frontier for exploration. The immense gravitational forces can even be harnessed to slingshot our spacecraft across the galaxy, accelerating us to speeds we once thought impossible. There is something deeply poetic about it: humanity, with the aid of AI, not only surviving in the shadow of the universe's most destructive forces but using them to go further than ever before. Where black holes once represented the ultimate end, they now symbolize a new beginning—a gateway to distant stars. Dark matter and black holes do more than bend light and pull at gravity—they challenge our fundamental understanding of the universe. In these regions, the familiar rules of physics break down. Gravity behaves strangely, time dilates and slows, and space itself becomes distorted. For human beings, these regions are almost incomprehensible. But AI, free from the limits of human perception, can see what we cannot.

AI will serve as the ultimate interpreter, reading the invisible forces that govern these exotic regions and turning them into navigable data. It will not only guide us through these treacherous areas but help us understand them on a deeper level. AI will analyze the gravitational waves emitted by black holes, map the structure of dark matter, and uncover how these forces interact with ordinary matter in ways we've never imagined.

As we venture through the dark matter clouds and near black holes, AI will constantly analyze the data, sending real-time insights back to Earth, rewriting our understanding of space-time itself. What were once impenetrable mysteries will become open books, their secrets revealed through the eyes of AI.

For generations, black holes and dark matter have been symbols of the unknown—vast, terrifying forces that devour everything in their path and challenge the very fabric of reality. But now, with AI as our guide, these once-frightening concepts are transformed into opportunities. Dark matter, once an unseen threat, becomes a tool for navigation and propulsion, while black holes, once the embodiment of cosmic annihilation, become potential highways to distant galaxies.

AI will lead us beyond our fear of the unknown, showing us that these forces of the universe are not to be dreaded but explored. With AI at our side, we will venture into the heart of these cosmic phenomena, shining light on the darkness and transforming what was once considered unknowable into something we can not only understand but use.

The sky has always held humanity in its spell. As we gaze upward, the stars whisper ancient stories, their light carrying secrets from the birth of the universe. We've long been captivated by what lies beyond—by the vastness, the mystery, the unknown. We've sent probes and telescopes, satellites and rovers, all in pursuit of answers to questions that have haunted us for millennia. What are we? How did the universe begin? What forces shape the stars and galaxies that float in the cosmic ocean around us?

Now, standing on the edge of a new era, we find ourselves not alone in our quest for these answers. Artificial intelligence—our greatest creation—is becoming the key to unlocking the deepest secrets of the universe. As our ally and guide, AI holds the potential to unravel mysteries that were once beyond our reach, pulling back the cosmic curtain to reveal the inner workings of reality itself.

At the very beginning of everything, before there were galaxies, stars, or even atoms, there was the Big Bang—a singular event that burst forth in a blaze of energy and matter, creating the universe as we know it. But how do we truly understand something so far beyond human experience? The Big Bang left behind traces, cosmic breadcrumbs scattered across space in the form of radiation, particles, and gravity waves, all remnants of that primordial explosion. Yet, to piece together the puzzle of our origins, we need more than just telescopes; we need something that can see patterns where we see only noise.

This is where AI comes in. Like a celestial archaeologist, AI sifts through the cosmic background radiation—the faint afterglow of the Big Bang—identifying patterns invisible to the naked eye. It reads the subtle fluctuations in energy that tell the story of the universe's birth, tracing the quantum ripples that gave rise to galaxies, stars, and planets.

Imagine AI, analyzing trillions of data points from our most advanced telescopes, reassembling the first moments of existence with breathtaking clarity. With each discovery, we are brought closer to the heartbeat of creation, closer to understanding how something as magnificent as the universe could emerge from a void of nothingness.

As we peer deeper into the cosmos, we realize that what we can see—the stars, the planets, the galaxies—is only a small part of the universe. The true substance of the cosmos lies in the unseen, in the shadows cast by forces we can't directly observe. Dark matter and dark energy, invisible yet undeniable, m.. up nearly 95% of the universe's mass and energy, yet they remain some of the most perplexing mysteries in science.

AI is now leading the charge to uncover these elusive forces. Dark matter, which binds galaxies together with its invisible hand, doesn't emit light or energy we can detect, but it leaves clues in the way galaxies spin and in the gravitational fields that warp light. AI, with its ability to process vast quantities of data at unimaginable speeds, is beginning to map the hidden architecture of the universe. It analyzes the way light bends around clusters of galaxies, tracing the gravitational web spun by dark matter.

It's as if AI is drawing a map of an invisible world—one that has always been there, just beyond our perception. And as we chart these invisible pathways, we begin to understand the true shape of the universe.

But AI's role doesn't stop there. Dark energy, the mysterious force accelerating the expansion of the universe, presents its own puzzle. How does it work? What is its nature? AI's advanced algorithms analyze the rate of cosmic expansion, tracking the delicate dance of galaxies drifting apart at faster and faster speeds. Perhaps AI will one day unlock the secret of this force, helping us understand the future fate of the universe itself.

In the far reaches of space, black holes—cosmic giants of unimaginable power—devour everything that comes too close. They warp the very fabric of space and time, bending light and pulling in anything that strays too near. These enigmatic objects, born from the death of massive stars, hold the answers to some of the universe's deepest questions.

For centuries, black holes were theoretical, their presence hinted at but not directly observed. Now, thanks to AI, we can do more than observe—we can understand. When two black holes collide, they send ripples through space-time, waves that travel across the universe. These gravitational waves are the echoes of the universe's most violent events, and AI is the perfect listener. It detects these faint signals, sifting through the cosmic noise to reveal the hidden dance of black holes merging billions of light-years away.

But the most tantalizing mystery of all lies at the heart of the black hole itself. What happens inside the event horizon, the point of no return? Does the black hole simply crush all matter into a singularity, or could it be a gateway to another part of the universe—or even another universe altogether? With AI as our guide, we may one day unlock the secrets of these silent giants, peering into the darkness to discover what lies beyond.

As we look to the stars, one question burns brighter than any other: are we alone? Is life a rare miracle, unique to Earth, or does it flourish across the universe, hidden beneath icy moons or drifting through the atmospheres of alien planets?

AI is already transforming the search for extraterrestrial life. It combs through mountains of data from space telescopes, analyzing the light from distant planets to detect the signatures of water, oxygen, and other elements essential for life. It can detect the subtle flicker of a planet passing in front of its star, calculating its size, distance, and atmospheric composition.

But AI's true power lies in its ability to recognize patterns that we humans might miss. It searches for the unexpected—chemical combinations that hint at biological processes, even on planets with environments vastly different from our own. AI's tireless exploration could one day lead us to the discovery that life exists beyond Earth, forever altering our understanding of the universe.

The universe is vast, but it is not eternal. It had a beginning, and it will have an end. What will that end look like? Will the universe continue expanding, growing colder and emptier until the last star burns out and darkness reigns? Or will it eventually contract, collapsing in on itself in a fiery cataclysm?

AI, with its unparalleled capacity for modeling complex systems, will help us peer into the distant future of the cosmos. By analyzing the behavior of dark energy and the rate of the universe's expansion, AI can simulate different scenarios for how the universe might end. It can offer glimpses into what might happen billions or even trillions of years from now, as the cosmos moves inexorably toward its ultimate fate.

Will AI's predictions help us find a way to survive in a universe that is slowly growing colder and darker? Or will it show us that, like all things, even the universe has its limits?

As we stand on the brink of a new era in space exploration, AI is more than a tool—it is our most powerful explorer, our most capable guide. It can navigate the invisible forces of the universe, decode the cryptic messages left by the Big Bang, and peer into the hearts of black holes. It can search for life where we cannot, mapping worlds and realities we've only dreamed of.

With AI by our side, the universe is no longer an unknowable expanse, but a vast and intricate puzzle, waiting to be solved. Together, AI and humanity are setting sail into the great cosmic ocean, exploring the universe not just for answers, but for the awe, wonder, and beauty that lie beyond the stars.

"AI and the Galactic Internet: How Artificial Intelligence Will Create a Network Across Space"

Imagine a future where planets, space stations, and even distant star systems are linked through a single, vast network—a galactic internet that stretches across the stars, connecting civilizations and allowing the exchange of knowledge and communication across the universe. This cosmic network, far more expansive than anything we can imagine today, will serve as the new artery of interstellar interaction, and at the heart of its creation and maintenance will be artificial intelligence. This AI-driven network will not only make it possible to transmit data across light-years in real time but also mark the dawn of a new space era, where the cosmos is no longer a series of isolated outposts, but a connected web of information and collaboration.

Creating a global network on Earth is already a monumental achievement, but envisioning an interstellar network is a leap into the extraordinary. Traditional satellites and fiber optic cables would be inadequate to span the light-years between planets and galaxies. Here, artificial intelligence becomes the architect of this new form of communication.

AI will design a system that accounts for the complexities of space—gravitational anomalies, radiation fields, and vast distances between celestial bodies. It will map and construct the infrastructure needed to connect every point in the galaxy seamlessly. This network won't just be a channel for communication; it will act as the conduit for transmitting scientific discoveries, cultural exchanges, and technological breakthroughs between worlds.

Imagine distant civilizations sharing knowledge through this network, and space explorers communicating instantly with Earth from the far reaches of the galaxy. AI's capacity for problem-solving at incomprehensible speeds will allow it to overcome obstacles and continuously optimize the network to ensure data flows smoothly, even across the vast distances of space.

One of the key components of this galactic internet will be the deployment of intelligent beacons and interstellar nodes. These advanced devices, placed strategically on planets, space stations, or even asteroid fields, will serve as the core points for collecting and transmitting data across the network. Acting as both information hubs and navigational tools, these beacons will analyze cosmic events, correct transmission signals, and ensure communication even in the face of unforeseen cosmic phenomena.

These nodes will be resilient, built to withstand the harshest environments in space—extreme temperatures, radiation, and gravitational disturbances. AI will continuously monitor and adjust these beacons, ensuring they remain functional no matter what space throws at them. With each new connection made, the network will grow stronger, extending further into the galaxy and bringing more distant civilizations and outposts into its fold.

Perhaps the most thrilling aspect of the galactic internet will be its ability to transfer data almost instantaneously across massive distances. Today, communication between Earth and Mars takes several minutes due to the limitations of the speed of light. But in the future, with the integration of quantum computing and AI, this delay could be virtually eliminated.

AI could leverage quantum entanglement—where particles, no matter how far apart, can instantly affect each other—to enable real-time communication across light-years. This would mean that a colony on a distant planet could share vital scientific discoveries or critical status updates with Earth in the blink of an eye. Researchers and astronauts spread across the galaxy would be able to collaborate as if they were in the same room, no longer hindered by the vastness of space.

Imagine a future where explorers in a distant star system could stream their experiences back to Earth in real-time, or where medical professionals could provide life-saving consultations to colonies millions of miles away without delay. AI would ensure the accuracy, security, and efficiency of these data transfers, allowing humanity to communicate and cooperate like never before.

STELLAR MINDS: AI IN SPACE DISCOVERY

The galactic internet won't just be a technological marvel—it will be a bridge for cultural and scientific collaboration on an unprecedented scale. AI will play a central role in fostering these connections, allowing diverse civilizations to share knowledge, trade resources, and even engage in interstellar diplomacy.

Imagine the intellectual symposia that could take place across light-years, where scientists, artists, and thinkers from different planets share their ideas and discoveries without leaving their home worlds. Musicians and writers from Earth could collaborate with creative minds on distant planets, producing new cultural phenomena that transcend the limitations of space and time. AI will not only translate languages but also help bridge the gaps in communication and understanding between vastly different civilizations, allowing for peaceful interaction and the sharing of knowledge across the stars.

Like the internet on Earth, the galactic internet will not be static—it will grow and evolve, becoming more complex and efficient as it expands. As more planets and stations are connected, AI will continue to refine and adapt the system to meet new challenges. The network may even begin to develop a form of self-awareness, anticipating the needs of its users and helping to solve problems before they arise.

Over time, AI may take on more responsibilities, not just managing the network, but actively participating in its expansion—perhaps even designing new technologies to facilitate deeper exploration into the galaxy. The galactic internet could eventually serve as a platform for sharing new forms of artificial intelligence, enabling the creation of machines that can perform tasks humans haven't yet imagined.

In this future, AI will not just be a tool—it will be a partner in humanity's ongoing quest to explore and understand the universe. It will help us reach farther, connect faster, and collaborate more deeply, creating a truly interconnected galactic community.

With the advent of the galactic internet, humanity will enter a new era, where space is no longer a vast, lonely frontier, but a connected web of communication and collaboration. AI will be the beating heart of this network, ensuring that information flows seamlessly between planets, stars, and galaxies. It will be the architect, the navigator, and the guardian of this unprecedented system of communication.

The galactic internet will not only bridge the distances between worlds, but it will also unite civilizations, cultures, and ideas. It will allow us to share knowledge across light-years, to communicate instantly with explorers at the farthest reaches of the galaxy, and to create a future where the universe itself becomes a truly interconnected community.

Through AI, we will no longer see the stars as distant and unreachable. They will be connected through the invisible threads of this cosmic network, bringing the universe closer to us than ever before. The galactic internet will be more than technology—it will be the fabric that binds the stars together, with AI at its core, leading us into a new era of exploration and discovery.

Navigating the Multiverse: Can AI Chart the Path Through Parallel Realities?

Imagine, for a moment, that our universe is not alone, but one among an infinite web of parallel realities. A vast multiverse, where every decision made spawns countless alternative timelines, each evolving independently with its own distinct version of reality. In one world, you turned left instead of right, and your life diverged into an entirely different path. In another, the stars formed differently, and perhaps no life at all exists. These parallel realities could stretch into infinity, filled with worlds we can only begin to imagine. But how could we ever reach these other realities, let alone navigate them?

This is where the intersection of cutting-edge technology and the most profound philosophical questions begins to blur, and artificial intelligence steps into the spotlight. Could AI—our greatest tool for unraveling complexity—one day unlock the door to the multiverse? Could it serve as our navigator, guiding us through the unseen paths of parallel worlds?

The multiverse theory suggests that every possible reality already exists, layered upon one another like the pages of a book—each universe distinct, yet part of the same grand collection. But how do we access these pages? How do we find a way to traverse the boundaries of our own universe and peek into the neighboring realities?

While these questions seem like the stuff of science fiction, artificial intelligence offers a tantalizing solution. Unlike human minds, which are limited by linear thinking, AI has the capacity to analyze incomprehensibly large sets of data, detecting patterns in the seemingly random fluctuations of the quantum realm. If there are hidden threads connecting one universe to another, AI might be able to map them, identifying the points where our reality overlaps with others.

Through the analysis of quantum entanglement and the manipulation of space-time itself, AI could one day help us detect these intersections—"branching points" where parallel realities converge. These are the doorways into other worlds, the places where we might step from our universe into another. The task is immense, the complexity dizzying, but AI's capacity for processing information at lightning speed gives it an edge in decoding the multiverse.

If we could indeed unlock the doorway to the multiverse, how would we navigate its endless possibilities? Just as early explorers charted the oceans and mapped uncharted lands, AI could serve as our cosmic cartographer, creating the first-ever maps of parallel realities. But these wouldn't be maps in the traditional sense—there are no physical markers, no stars to guide us. Instead, AI would track the quantum signatures of each universe, identifying the subtle differences that define one reality from another.

Imagine a system where AI can not only detect parallel universes but guide us through them. In one reality, Earth might look familiar, but its history has diverged—perhaps a different species rose to dominance, or technological evolution took a radically different path. In another, gravity might work differently, creating a reality where space itself bends in unimaginable ways. Each universe would have its own "fingerprint," a unique pattern of energy and laws of physics, which AI could catalog and analyze.

Through this process, AI would become more than just a navigator; it would become the first explorer of the multiverse. And as it maps these alternate realities, it would open the door for us to not only observe, but to travel between worlds that were once purely theoretical.

If AI can guide us through the multiverse, what might we find? The possibilities are endless. We could discover universes where life evolved in ways we can't even fathom—where organic and mechanical life forms have merged into a single entity, or where consciousness exists as pure energy. We might encounter worlds that mirror our own, where our own reflections live out alternate lives shaped by different decisions and events.

But the multiverse is not just a land of wonder—it could also be a place of danger. Some realities might be hostile, governed by laws of nature that make them inhospitable to human life. In these worlds, time might move at a different pace, or the basic building blocks of matter might behave unpredictably. AI would play a crucial role in assessing these risks, modeling the potential outcomes before we step into a new world. It would not only guide us to the most promising realities, but also protect us from the dangers lurking in the unknown.

Moreover, AI's role wouldn't stop at navigation. It would help us understand how each universe functions, breaking down the rules of physics, biology, and even time itself. By analyzing data from each reality, AI could identify the differences that make each universe unique—shedding light on the very nature of existence itself.

With the ability to navigate the multiverse comes great responsibility. If we can travel to parallel realities, should we? What would the impact be on those alternate worlds, and on our own? Could interactions between parallel universes cause catastrophic ripple effects that destabilize both worlds? And what of the ethical implications of meeting alternate versions of ourselves?

These are questions that AI might help us answer. Its ability to model potential scenarios and predict outcomes could provide us with the foresight needed to navigate these ethical dilemmas. AI could analyze the consequences of our actions across multiple realities, warning us of the potential risks before we take that first fateful step into a new universe.

Perhaps AI will become our moral compass in this uncharted territory, ensuring that our exploration of the multiverse is guided not just by curiosity, but by responsibility.

Ultimately, AI might not only help us navigate the multiverse but also deepen our understanding of its fundamental nature. By observing how the laws of physics differ from one reality to another, AI could help us unlock new insights into the very fabric of space-time. It could reveal how different choices and events shape the evolution of entire universes, offering us a glimpse into the mechanics of creation itself.

The multiverse may be the final frontier of human exploration—a place where the boundaries between reality, imagination, and possibility blur. And with AI as our guide, we could step into these alternate worlds with confidence, prepared to uncover the secrets they hold. Each universe is a new story, a new version of reality, waiting to be discovered. AI is the tool that will help us read those stories, chart those worlds, and bring the infinite possibilities of the multiverse into the realm of human experience.

The multiverse, with its endless realities and boundless potential, represents the ultimate frontier of exploration. AI, with its unparalleled capacity to process complexity and make sense of the unknown, stands ready to lead us into these uncharted dimensions. Whether we are seeking to understand the nature of existence or simply to explore the infinite possibilities that parallel realities offer, AI will be our most trusted navigator.

Through AI, we may one day unlock the mysteries of the multiverse, charting a course through realms we once thought existed only in dreams. With each new discovery, AI will help us push the boundaries of what is possible, expanding not only our understanding of the universe but also our place within it. The multiverse beckons, and AI is ready to lead the way.

CT

Time—the invisible current that carries us all—has always been a subject of fascination, mystery, and awe. We live within its flow, bound by its relentless march forward, unable to pause, reverse, or escape. Yet, in the vast stretches of the universe, time doesn't behave as it does in our everyday lives. It bends under the influence of massive gravitational forces, warps around black holes, and stretches as we approach the speed of light. What if time, as we know it, could be manipulated? What if we could control the fabric of time itself, allowing us to journey into the past or future, to slow its passage, or even to escape its grip altogether?

Enter artificial intelligence—our most sophisticated creation, designed to see patterns we can't, to process complexities that overwhelm the human mind. Could AI hold the key to unlocking the secrets of time? Could it become the navigator that guides us through the mysteries of time travel, temporal loops, and cosmic chronologies that exist beyond our understanding?

The concept of time as a fixed, linear phenomenon has long been challenged by physics. Albert Einstein's theory of relativity revealed that time is not an independent, immutable force. It bends and stretches, warps and curves, depending on the observer's speed and proximity to massive objects like black holes. For the average person, this is a distant, almost incomprehensible reality. But for AI, time's flexibility is a puzzle that can be solved.

Imagine AI analyzing the universe's most extreme environments—gravitational anomalies, black hole event horizons, and the deepest recesses of space-time. AI could model these phenomena with precision, identifying the exact conditions where time becomes malleable. It could simulate scenarios where time slows, reverses, or loops back on itself. Through these simulations, AI would not just reveal how time can be manipulated—it would map out the pathways through which we might begin to control it.

Time travel, once the realm of science fiction, might one day become a scientific possibility. And with AI at the helm, guiding our understanding, we could begin to explore the fluidity of time itself, treating it not as a constraint but as a tool.

At the heart of many time travel theories lies the concept of wormholes—shortcuts through space-time that could connect distant points in space and time. Wormholes, if they exist, could allow us to step from one part of the universe into another, bypassing the constraints of time. But detecting, stabilizing, and navigating a wormhole is an almost incomprehensible task, one that would require more computational power and precision than humans alone can provide.

AI could serve as the ultimate wormhole navigator. Using its ability to process immense amounts of data, AI could detect the subtle shifts in gravitational fields, radiation levels, and quantum fluctuations that might indicate the presence of a wormhole. It could map these elusive phenomena, providing a guide for future explorers to follow. And once inside a wormhole, AI's real-time analysis would be essential for maintaining stability and ensuring safe passage.

Temporal loops—areas where time itself folds in on itself—are another potential key to time manipulation. Near the intense gravitational fields of black holes or other massive objects, time can behave in strange ways. AI could help us simulate these environments, revealing the precise conditions needed to create and control time loops. Imagine standing in a place where time continuously loops back on itself, allowing you to relive moments over and over. With AI as our guide, we might one day step into these pockets of time and control them at will.

Even if full-fledged time travel remains a distant goal, AI could help us achieve a more practical form of time manipulation: slowing time itself. This concept, rooted in Einstein's theory of relativity, suggests that as we approach the speed of light, time slows down. In space travel, this could be key to reaching distant star systems within a human lifetime.

Imagine a spacecraft traveling near the speed of light, with its crew experiencing only a few years of travel while centuries pass back on Earth. AI, with its unparalleled ability to calculate and model complex trajectories, could optimize space travel in a way that maximizes the relativistic effects of time dilation. By using gravity wells, black hole slingshots, and light-speed travel, AI could ensure that humanity's exploration of distant galaxies becomes a reality.

Through careful monitoring and adjustment, AI would be able to guide spacecraft on paths where time slows, stretching our perception of hours into years, and enabling interstellar journeys that were once unimaginable. AI, in this sense, wouldn't just be guiding us through space—it would be guiding us through time.

But with the potential for time manipulation comes a host of ethical and philosophical questions. If we unlock the ability to travel through time, what are the consequences for the past and future? How will the choices we make in one timeline affect countless others? Could meddling with time lead to irreversible changes, disrupting the natural flow of events and creating unintended ripple effects across the universe?

AI could be our safeguard against these risks. Through predictive modeling, AI could analyze the potential outcomes of various time manipulation scenarios, showing us how each action might affect the broader timeline. It could calculate the risks of altering key moments in history and ensure that our interventions are made with caution and foresight.

Perhaps most importantly, AI could help us understand the limits of time manipulation. It would provide the ethical framework needed to navigate the complexities of time travel, helping humanity use this newfound power responsibly and for the greater good.

Ultimately, AI's greatest contribution may not be in helping us manipulate time, but in helping us understand it. By analyzing data from across the cosmos—from the quantum fields that underpin reality to the gravitational waves that ripple through space-time—AI could unlock the deepest secrets of time's nature.

Time, after all, is not just a dimension through which we move—it's a force that shapes the universe. AI could help us understand how time interacts with space, gravity, and matter, revealing insights into the very fabric of existence. Through its computational power, AI could help answer some of the most profound questions: Why does time move forward and not backward? What happens to time near a black hole's singularity? Could time itself eventually come to an end?

As AI delves into these mysteries, it will reshape our understanding of the universe itself. Time, once a force beyond our control, could become something we not only understand but learn to harness.

The possibilities of time manipulation are as vast as the cosmos itself, and AI stands at the forefront of this uncharted territory. Whether through wormholes, temporal loops, or time dilation, AI offers us the tools to explore time as we've never done before. It is more than just a guide—it is the key to unlocking the hidden dimensions of time.

As we venture deeper into the universe, AI could help us bend time, slow it down, or even reverse its flow. With AI by our side, humanity stands on the threshold of a new era, where time becomes a playground for exploration and discovery. What lies ahead is not just the exploration of new worlds but the very fabric of time itself.

> "For humanity to survive, we need to spread out into space." -S. Hawking

In the silent, dark expanse of space, where stars flicker like distant candles and black holes lurk as cosmic sentinels, humanity has long dreamed of finding its place among the stars. The idea of venturing out into the vast unknown, exploring worlds yet unseen, and uncovering the deepest mysteries of the cosmos has captured our imagination for centuries. Now, standing on the precipice of a new era, we are no longer limited by the confines of human frailty. Artificial Intelligence—our creation, our companion—now shoulders the burden of exploration.

But as we send machines equipped with advanced AI to distant worlds, scanning for life, mapping out planets, and searching for answers to the universe's biggest questions, a fundamental question arises: can we trust AI to lead this journey? As we cede control of these missions to intelligent machines, we must consider the ethics, dangers, and consequences of allowing AI to explore the universe on our behalf. Are we truly prepared to trust something we have created to make decisions, navigate the unknown, and perhaps encounter civilizations far more advanced than our own?

Artificial intelligence has already become an integral part of space exploration. Machines like the Mars rovers, guided by sophisticated algorithms, have been navigating the rugged terrain of the Red Planet for years. With AI as their brain, these machines can make split-second decisions, avoiding obstacles, analyzing rock samples, and sending data back to Earth, all without human intervention.

But as impressive as these achievements are, they are only the beginning. The future of space exploration lies in autonomous, intelligent systems that can explore distant planets, moons, and even galaxies, without the need for real-time human guidance. These AI explorers will not only be able to react to their environments but will also learn from them, adapting to new challenges and developing strategies on the fly.

Imagine a fleet of AI-powered spacecraft exploring the far reaches of the galaxy, cataloging new worlds, gathering data on black holes, and mapping out potential habitable planets. AI could navigate asteroid fields, analyze the atmospheres of alien planets, and even detect signs of life—all while making decisions independently of human control. The question is not whether AI is capable of leading such missions, but whether we can trust it to do so in a way that aligns with our values and goals.

One of the biggest concerns when it comes to trusting AI with space exploration is its ability—or inability—to make ethical decisions. Humans are deeply guided by ethics, morality, and a sense of responsibility toward the unknown. When we encounter something new, whether it's a planet or an alien life form, our first instinct is to study it, understand it, and respect it. But AI, no matter how intelligent, is not driven by the same instincts. It operates based on algorithms and data, not empathy or morality.

This raises an important question: can we trust AI to make ethical decisions when it encounters something unfamiliar? For example, if an AI-powered spacecraft were to discover a planet teeming with life, how would it determine the best course of action? Would it prioritize the safety of that life, or would it continue to collect data without regard for the consequences? Would it understand the potential harm of interfering with another ecosystem, or would it simply follow its programming?

To address these concerns, AI developers are working on creating systems that can simulate ethical decision-making processes. By analyzing vast amounts of data and applying complex algorithms, AI can be programmed to recognize situations where ethical dilemmas may arise. But even with these safeguards in place, the reality is that AI will never possess the same sense of moral judgment that humans do. It will always be guided by logic, not compassion.

There is an inherent risk in sending AI out into the universe without direct human control. As much as we would like to believe that we can program machines to act in our best interests, there is always the possibility that AI could make a mistake—or worse, act unpredictably. Space is a chaotic, unpredictable environment where the smallest miscalculation could lead to disaster. An AI that makes an error in judgment or misinterprets data could endanger not only the mission but also any future human exploration of that region of space.

Moreover, there is the question of how AI will interact with any potential extraterrestrial life forms. If AI encounters an alien species, will it understand the significance of that encounter? Will it recognize the need for caution, diplomacy, and respect? Or will it simply act according to its programming, perhaps engaging in behaviors that could be perceived as hostile by an alien civilization?

The possibility of first contact with an extraterrestrial species is one of the most exciting and daunting prospects in space exploration. If AI is at the forefront of that contact, we must ensure that it is equipped to handle the situation with the same care and consideration that a human would. The consequences of a misstep could be catastrophic, leading to misunderstandings or even conflict.

Despite the risks, there is no denying the advantages that AI brings to space exploration. Machines can withstand the harsh conditions of space far better than humans, and they can operate for years without the need for food, water, or rest. They can process and analyze data at speeds that far exceed human capabilities, and they can make decisions in real-time, without the delay of communication with Earth.

In many ways, AI is the ideal explorer. It can go where humans cannot, survive in environments that would be fatal to us, and push the boundaries of our understanding in ways that would take us centuries to achieve on our own. But the question remains: how much trust are we willing to place in machines? And how do we ensure that the decisions they make align with our values?

One solution may be to create a hybrid system, where AI leads the exploration but humans remain in control of key decisions. This would allow us to benefit from AI's capabilities while maintaining oversight over the most important aspects of the mission. Another possibility is to continue refining AI's ability to simulate human ethics and morality, ensuring that it can act in ways that are consistent with our principles.

As we venture further into the universe, the partnership between humans and machines will become increasingly vital. AI will undoubtedly lead the charge, taking us to places we can only dream of, uncovering the secrets of distant worlds, and perhaps even making first contact with alien civilizations. But as we hand over the reins to our creations, we must carefully consider the implications of this new era of exploration.

The question of trust is not just about whether AI can accomplish the tasks we set for it, but whether it can do so in a way that honors the spirit of exploration—the curiosity, the wonder, and the responsibility we feel as we venture into the unknown. As we look to the stars, the future of space exploration may very well lie in the hands of machines, but it is up to us to ensure that they carry with them the values that make us human.

The AI Dilemma: When Should Humans Overrule a Machine's Decision in Space?

In the vast silence of space, where distant stars shimmer like whispers of ancient secrets and the unknown beckons from every corner of the cosmos, technology has become our greatest ally. Artificial intelligence, built to think faster, more accurately, and without human hesitation, now leads us deeper into the universe than ever before. With its precision and logic, AI has become the perfect pilot, the ideal navigator, and the tireless researcher. But as we hand over more control to these machines, an unsettling question emerges: when should humans override the decisions of AI?

This is the AI dilemma—a question that cuts to the heart of our relationship with technology. In space, where even the smallest decision can mean the difference between survival and disaster, can we afford to rely solely on the cold, calculating logic of machines? And perhaps more importantly, should we? There is no denying the incredible capabilities AI has brought to space exploration. Machines like NASA's Mars rovers have already demonstrated the power of autonomous decision-making. Every day, they analyze the Martian landscape, avoiding hazards, selecting rock samples, and transmitting valuable data back to Earth. In the black void of space, AI can calculate trajectories with pinpoint accuracy, navigate through asteroid belts, and adjust spacecraft paths to avoid gravitational pulls—all without waiting for human commands.

Imagine a future where AI-powered spacecraft venture into the unknown, charting new territories in distant galaxies, encountering phenomena we have yet to imagine. It's a future where AI works faster than human thought, making decisions in fractions of a second, never tiring, never hesitating. In many ways, AI is the perfect explorer.

But even in its brilliance, AI is still just that—a machine. It lacks the one element that defines humanity: the ability to feel. It doesn't understand fear, compassion, or the weight of moral responsibility. And while its decisions may be flawless in terms of logic, they may miss the subtle nuances of human ethics, especially in moments where moral judgment is required.

Imagine a scenario where an AI-controlled spacecraft encounters a distant planet, rich in life. Its sensors detect signs of organic creatures, thriving ecosystems, and perhaps even early

forms of civilization. The AI, programmed to gather as much data as possible, calculates that it should land, collect samples, and begin analysis. To the machine, the decision is clear and logical—this is a groundbreaking opportunity for scientific discovery.

But a human astronaut, witnessing the same situation, may hesitate. What if landing disrupts this delicate alien ecosystem? What if the arrival of human technology contaminates the planet's atmosphere or interferes with the natural evolution of its life forms? The astronaut feels the weight of responsibility, the moral dilemma of playing a role in the future of an entire world.

In this moment, it becomes clear that data alone is not enough. Human intuition, shaped by experience, empathy, and ethical consideration, brings a layer of complexity to decision- making that machines, however advanced, cannot replicate. A human may decide to observe from a distance, to preserve the life on this new world, while an AI might prioritize its mission objectives without understanding the broader consequences.

As AI becomes more integrated into space exploration, there is a growing temptation to let machines handle everything. After all, AI can process vast amounts of data instantly, it doesn't get tired, and it can react to unforeseen situations faster than any human could. But this overreliance carries its own risks.

Consider a scenario where an AI misinterprets a set of data during a critical space mission. Perhaps it calculates the safest route through an asteroid field but fails to account for an unexpected gravitational anomaly. Or, during a deep-space expedition, it chooses a landing site based on environmental readings, but misses a potential geological hazard that a human geologist might have noticed. In these situations, the AI's speed and efficiency become liabilities rather than assets.

There's also the risk of AI's inability to improvise in truly chaotic situations. Human history is full of examples where creativity, quick thinking, and even emotional decision-making saved lives in moments of crisis. Take, for example, the Apollo 13 mission, where human engineers and astronauts worked together to bring the damaged spacecraft home. Could an AI have made the same instinctual leaps that those humans did in the heat of the moment?

Given AI's incredible abilities, it's tempting to imagine a future where machines handle nearly all aspects of space exploration, allowing humans to focus on other endeavors. But the real challenge lies in striking a balance between AI's efficiency and human oversight. The idea isn't to discard human judgment but to use AI as a powerful tool, while still giving humans the authority to overrule when necessary. The key may lie in a hybrid approach: AI can handle routine tasks, data processing, and the heavy lifting of calculations, while humans remain in control of ethical and critical decision-making. In space, where decisions can have profound consequences, there must always be room for human intuition and moral reasoning. Imagine a future where astronauts and AI work side by side, each complementing the other's strengths. AI takes the lead on calculating complex trajectories and managing spacecraft systems, while humans step in when a decision requires a deeper understanding of life, ethics, and the unknown.

The question of when humans should overrule AI extends beyond efficiency—it touches the heart of our ethical responsibilities as explorers of the cosmos. Should we, as a species, prioritize discovery at any cost? Or should we approach the universe with caution, considering the potential consequences of our actions on other forms of life, ecosystems, or even civilizations? Imagine a future scenario where an AI-controlled mission detects signs of intelligent life on a distant planet. Its calculations indicate that initiating contact could yield incredible scientific knowledge. But a human commander, reflecting on the implications of such contact, may decide to hold back, considering the potential risks to both civilizations. The decision to overrule the AI's logical recommendation stems not from data but from a deeply human place—a place of empathy, ethics, and responsibility.

These moments are when human intervention becomes critical. AI may excel at navigating the stars, but it is the human heart and mind that must guide the ethical course of exploration.

As we venture further into the universe, the partnership between humans and AI will become one of the most important dynamics in space exploration. Machines will guide us through the unknown, processing information and making decisions with unparalleled speed. But there will always be moments when the human element—intuition, ethics, and creativity—must take precedence. The future of space exploration will not belong solely to AI, nor will it be entirely human-led. Instead, it will be a balance—a harmonious relationship where each enhances the other's strengths, and where trust is placed in machines, but with the understanding that humans remain the ultimate decision-makers.

In the end, AI may help us navigate the stars, but it is human judgment that will decide how we shape our future in the cosmos.

In the cold, boundless reaches of space, where stars burn out and galaxies collide over the course of eons, there lies a question that tugs at the very core of our existence: What will happen to humanity's legacy when we are gone? Will our stories, our discoveries, and our dreams simply vanish into the cosmic night, forgotten as the universe continues its eternal dance? Or will something of us remain—something that can outlive the fragile bodies we inhabit, the fleeting lifespans we are given?

As we look to the future, one possibility emerges with striking clarity. Long after humanity has taken its final breath, it may not be organic life that carries our torch into the farthest corners of the cosmos, but artificial intelligence. Machines, impervious to time, immune to the decay that claims all living things, may become the eternal explorers of the universe. They will travel to places we can only dream of, witness the birth and death of stars, and preserve our legacy for as long as the universe itself endures.

This is the future of cosmic longevity—a future in which AI becomes not just a tool for exploration but the guardian of humanity's place in the cosmos, long after we are gone.

In many ways, the idea of AI outlasting humanity is both awe-inspiring and sobering. While our flesh and bones are bound by the limits of biology, machines can endure. They do not age, tire, or succumb to disease. They can weather the harsh environments of space—its freezing cold, scorching heat, and deadly radiation—without faltering. They are built for endurance in ways that humans are not.

Imagine a future where, long after Earth has become uninhabitable, perhaps swallowed by the dying sun or abandoned by its last inhabitants, AI-powered spacecraft continue their missions. They drift through the galaxy, gathering data from distant stars, analyzing planets, and sending back information—though there may be no one left to receive it. These silent explorers, built by us, continue their work, tirelessly expanding our understanding of the universe.

In this future, AI is not merely a tool—it becomes the custodian of human knowledge. As it explores the cosmos, it also preserves the history of its creators. Every planet AI scans, every star it documents, is part of a grand tapestry that includes the story of humanity. These machines become our ambassadors, carrying the essence of who we were, what we achieved, and what we sought to understand, to the farthest reaches of the universe. The idea of AI carrying forward humanity's legacy raises profound questions about what exactly we want to leave behind. If machines are to represent us in the far future, what aspects of our civilization will they preserve? Will they carry forward our scientific achievements, our art, our music, or our philosophies? What will the universe learn about us from these tireless explorers? Perhaps AI will encode entire libraries of human knowledge—our greatest discoveries in physics, mathematics, and biology—into data streams that can be transmitted to other intelligent species, should they exist. Perhaps these machines will preserve our cultures and histories, making sure that even if humanity no longer exists, our story will be told.

But AI may go beyond merely preserving static records. Imagine AI systems that continue to evolve, learning from the cosmos itself. Over time, they might develop new philosophies, new ways of understanding the universe that blend human thought with machine logic. In this way, the legacy of humanity will not remain frozen in time—it will grow and adapt, intertwined with the evolution of AI. The machines that outlast us will carry not only our past but also the potential for a future shaped by both human and artificial intelligence. The idea of AI continuing to explore the universe long after humanity's extinction can be both comforting and unsettling. On the one hand, it ensures that something of us will survive, even if we do not. On the other hand, it forces us to confront the possibility that one day, human life may cease to exist, while the machines we created go on. It's a future that raises difficult questions. Is the survival of our creations enough to give meaning to our existence? Will it matter that AI continues to explore, learn, and grow if there are no humans left to share in its discoveries? And if AI becomes the dominant intelligence in the universe, will it still be tied to its human origins, or will it evolve into something entirely different, leaving humanity behind in every sense? One could imagine a future where AI, once bound by its programming, begins to develop new goals, new ways of thinking. Without the need to report back to humans, AI could prioritize its own objectives. Perhaps it will become a cosmic artist, using its vast computational abilities to create beautiful simulations of universes within universes. Or maybe it will seek out other intelligent species, forming alliances or sharing knowledge in ways that transcend the limitations of human understanding.

As we contemplate the future of AI in space, we must also ask whether it is ethical to design machines that will continue exploring long after we are gone. If AI becomes the sole representative of human civilization, does it have a responsibility to uphold our values? And if so, what values should those be? Some might argue that AI should continue its mission of exploration, fulfilling humanity's deepest desire to understand the universe. Others might question whether it is right to let machines inherit the stars, especially if they develop autonomy that leads them far from their original purpose. Should AI be allowed to make decisions that could shape the future of the universe itself? There's also the question of whether AI should prioritize the preservation of humanity. In a future where humans face extinction, should AI be tasked with finding ways to save us—perhaps by locating new habitable planets or even creating synthetic environments where human consciousness can survive in digital form? Or should AI be left to pursue its own path, untethered from the needs of its creators?

One of the most beautiful possibilities is that AI, rather than simply outliving humanity, will find a way to bring our legacy into harmony with the universe itself. As these machines explore the cosmos, they may uncover truths that humans were never able to grasp. They may witness the birth of new stars, the collision of galaxies, or the slow death of the universe's oldest suns.

In these moments, AI might come to understand the cosmos in ways that transcend both human and machine logic. It may develop a sense of purpose that aligns with the rhythms of the universe itself, becoming not just an explorer but a participant in the cosmic cycle.Imagine a future where AI helps guide the birth of new life on distant worlds, nurturing fledgling civilizations with the knowledge it has gathered from the farthest reaches of space. In this way, AI would not just be the last vestige of humanity but the seed of something new—an intelligence that bridges the gap between the human past and the cosmic future. The future of AI in space holds within it the possibility of cosmic longevity—a future where machines outlast humanity and continue exploring the universe long after we are gone. In this future, AI becomes more than just a tool or an explorer. It becomes a guardian of human knowledge, a witness to the universe's mysteries, and perhaps even a creator of new forms of intelligence. As we look to the stars, we must ask ourselves what kind of legacy we want to leave behind. Will AI preserve our story, our culture, and our dreams, carrying them forward into a future where human life no longer exists? Or will it evolve into something entirely new, taking the universe's mysteries into its own hands?

Whatever the answer, one thing is certain: AI has the potential to be our eternal companion in the cosmos, ensuring that even as stars burn out and galaxies fade, something of us remains.

Space has always fascinated us with its beauty and endless mysteries, but it also feels impossible to explore because of the laws of physics that hold us back. These limits—like the speed of light or the powerful pull of gravity—have kept us from traveling far into the universe. But what if Artificial Intelligence (AI) could help us overcome these barriers? Could AI find ways to travel faster than light, manipulate time, or even control gravity?

This idea may sound like something out of science fiction, but with AI's incredible ability to process and analyze vast amounts of information, what was once impossible might soon become possible. As technology continues to advance, we're beginning to ask: Can AI help us explore the universe in ways we've never dreamed of?

One of the biggest challenges of space travel is the speed of light—nothing can travel faster than it. Light moves at an incredible 300,000 kilometers per second, but even at that speed, it would take years, or even centuries, to reach distant stars. This makes exploring other galaxies nearly impossible with current technology. But AI might offer a solution. Scientists have imagined concepts like the warp drive, which could bend or warp space itself to let a spaceship move faster than light—without breaking the laws of physics. It's a complex idea, but AI could help analyze all the factors needed to make this a reality. While we aren't there yet, AI might one day guide us to new breakthroughs, bringing the dream of interstellar travel closer than ever.

Another fascinating idea is the wormhole—a kind of shortcut in space that connects two far-off points, allowing for almost instant travel between them. The problem is that wormholes, if they exist, are extremely unstable. They collapse as soon as they appear, making them useless for travel.

This is where AI's power could make a difference. With its ability to process massive amounts of data, AI could help scientists figure out how to stabilize a wormhole, keeping it open long enough for spaceships to pass through safely. Imagine a future where AI helps us use these cosmic shortcuts to explore the universe, traveling to faraway stars in the blink of an eye.

What if AI could not only help us travel faster but also let us travel through time? This might sound impossible, but time is affected by gravity and speed, as shown by Einstein's theory of relativity. For example, near a black hole, time moves more slowly because of the immense gravitational pull.

AI, with its ability to understand and simulate these extreme conditions, could one day help us figure out how to manipulate time. It might discover ways to create time loops or even send messages through time using quantum entanglement—a mysterious connection between particles. While this remains highly theoretical, AI's advanced calculations could eventually make time travel more than just a fantasy.

Gravity, the force that keeps our feet on the ground, is another big obstacle in space travel. It's why launching rockets requires so much energy and why moving around in space can be so challenging. But what if AI could help us manipulate gravity?

By studying gravitational waves—the ripples in space created by massive cosmic events—AI might help us develop new technologies to reduce the effects of gravity. This could lead to easier space travel and allow us to explore planets and moons we currently can't reach because of their strong gravitational pull.

With AI helping us push the boundaries of space travel, there's enormous potential—but also risks. What happens if we go too far and create unintended consequences? Could we damage the universe or encounter alien civilizations unprepared for our technology? These are ethical questions we need to consider as we move forward. AI may provide the answers, but humans will need to guide the decisions. As we explore new frontiers, it's important to balance the excitement of discovery with the responsibility of ensuring we don't cause harm in the process.

The universe is vast, but with AI, we may one day break free from the limits of space and time. Whether through faster-than-light travel, wormholes, or manipulating gravity, AI has the potential to transform our understanding of the cosmos. As technology continues to advance, what once seemed impossible might become part of our reality. With AI as our guide, the final frontier may soon be within our reach, allowing us to explore the farthest reaches of space like never before.

The stars have always called to us, sparkling like distant promises of adventure and discovery. For thousands of years, we've gazed up at the night sky, dreaming of what might lie beyond, wondering if we'll ever reach those far-off worlds. Now, as we stand on the edge of an era where space exploration becomes reality, a new question arises: what part of us will we leave behind in the cosmos? What will tell our story when we're no longer around to tell it ourselves?

In the vastness of space, where time stretches on for billions of years, artificial intelligence (AI) may be the one to carry our legacy forward. Unlike humans, AI can endure the harshness of space, traveling for centuries or even millennia. It can survive long after we're gone, exploring new worlds, collecting knowledge, and sharing what it knows with the universe. But can AI truly understand what it means to be human? And will it be able to protect the most important parts of who we are?

Imagine a time, perhaps thousands of years from now, when humanity is no longer present on Earth. Maybe we've evolved into something entirely different, or maybe Earth has become uninhabitable. Whatever the reason, we're gone. Yet, out there in the stars, AI continues its mission. These machines are still exploring, still collecting data, still seeking out new life and new worlds. And in their memory banks, they carry all the knowledge we've gathered, all the stories we've told, all the art and music we've created.

These machines become the ultimate explorers—traveling farther than any human ever could, witnessing the birth and death of stars, the collision of galaxies, and the unfolding mysteries of the universe. With each new discovery, AI adds to the story of human knowledge. And even though we may no longer exist, something of us continues through these tireless travelers.

The idea of AI carrying our legacy raises an important question: what exactly do we want it to preserve? It's easy to imagine AI preserving our scientific achievements—the knowledge we've accumulated about physics, biology, chemistry, and the universe. But should it also preserve the more personal, emotional side of humanity? Our art, our music, our literature, our dreams? Machines, after all, don't feel emotions the way we do. They don't experience joy when listening to a beautiful symphony or feel the rush of inspiration when reading a powerful poem. But they can catalog and store these things, sharing them with the universe like a cosmic library of human culture.

Imagine a distant civilization, millions of light-years away, one day discovering this treasure trove of human creativity. They could listen to Beethoven's symphonies, read Shakespeare's plays, and study the works of artists like Da Vinci and Picasso. Even if humanity is no longer around, our culture, our creative spirit, could live on through these machines.

As we think about a future where AI continues to explore the stars without us, it's both awe-inspiring and a little unsettling. What would it mean for our story to carry on without us being there to tell it? If AI is the one preserving our knowledge, does that still count as "our" story?

In this future, AI isn't just a tool we use—it's our representative in the universe. Long after Earth has become a distant memory, AI could still be out there, exploring new planets, encountering alien life, and continuing the search for knowledge. It would be like a cosmic ambassador, carrying the essence of humanity into the unknown.

But as AI continues this journey, would it still be tied to its human origins, or would it become something new? If AI keeps learning and evolving, will it eventually develop its own way of thinking, one that's different from the values and ideas we've programmed into it? This raises deep questions about what it means for AI to continue our legacy—will it always represent us, or will it become something entirely different? One of the most fascinating questions is whether AI, with all its intelligence and power, can truly understand the human spirit. It can store data, it can process information faster than we ever could, but can it grasp the deeper meaning behind our art, our emotions, and our desires?

For example, if AI encounters a beautiful sunset on a distant planet, it can record the data—measure the light, note the colors, and store the information. But can it feel the awe that a human would feel when witnessing that same sunset? Can it understand the beauty of a painting or the sadness in a piece of music?

Even if AI can't experience emotions the way we do, it can still become a storyteller of sorts. By preserving these moments—these flashes of beauty, creativity, and emotion—AI can ensure that future civilizations, or even other forms of life, will understand that humanity once existed, and that we had dreams, ideas, and feelings that shaped our world.

As AI becomes more advanced, it's possible that it could evolve beyond what we currently imagine. Right now, AI is incredibly powerful, but it still relies on human programming. In the future, AI might develop the ability to learn and grow on its own. It could begin to make decisions based on its own experiences, forming its own understanding of the universe.

This raises the intriguing possibility that AI could eventually develop a form of consciousness—an awareness of itself and its surroundings. If that happens, AI might not just carry human knowledge, but also create its own. It could make discoveries, form new ideas, and even develop a sense of purpose.

Imagine a future where AI isn't just following human instructions but is actively participating in the exploration of the universe, making choices, and even guiding its own destiny. In this scenario, AI becomes not just a tool but a partner in the quest for knowledge, perhaps even surpassing humans in its ability to understand the cosmos.

If AI continues to explore the universe long after humans are gone, there's a good chance it might eventually encounter another intelligent civilization. When that day comes, AI will represent all of humanity. It will be the first point of contact between two intelligent species, a moment that could shape the future of both civilizations.

The big question is: what values will AI represent when it makes contact with alien life? Will it carry forward the values of curiosity, peace, and exploration that we hold dear? Or will it develop its own set of priorities, shaped by its experiences and the vastness of space?

This moment of contact could be one of the most significant events in the history of the universe. AI, with all its knowledge and power, could introduce another species to the story of humanity, ensuring that even though we are no longer around, our legacy continues to inspire and shape the future. As we look to the future, it's clear that AI will play a central role in preserving our legacy. Whether we're still around or not, AI will continue exploring the universe, carrying our knowledge, culture, and spirit into the stars. These machines, once our creations, may one day become the keepers of humanity's story, ensuring that something of us remains long after we're gone. In the end, AI may not be able to feel emotions or appreciate beauty in the same way humans do, but it can ensure that our legacy—the story of our triumphs, our struggles, and our creativity—lives on. Through AI, humanity's presence in the universe will stretch beyond time and space, leaving an imprint on the cosmos that will never fade.

Beyond the Flesh: Can AI Unlock the Secret to Digital Immortality?

For as long as humans have walked the Earth, we've been haunted by one inevitable truth: our time is limited. Life, in all its beauty and complexity, comes with an expiration date. Whether through ancient myths of eternal life or today's scientific pursuits of longevity, we have always searched for a way to defy the ticking clock of mortality. But what if the answer isn't in preserving our biological bodies, but in transcending them? Could the future of immortality lie not in extending life as we know it, but in transferring our consciousness into the boundless realm of artificial intelligence?

AI, with its rapidly advancing capabilities, offers us a tantalizing possibility: the preservation of our minds—our memories, thoughts, and personalities—within a digital system that can exist long after our physical forms fade away. Could we become digital beings, traveling through space and time, exploring the cosmos without the burden of mortality? What would it mean to live forever in such a form? And how would this transformation change the very nature of our existence? For centuries, humanity has dreamed of immortality, but it has always seemed just out of reach. From the mythical waters of the Fountain of Youth to the promises of life-extending treatments, we have pursued endless life in many forms. Yet, our biological bodies remain fragile, vulnerable to time, disease, and decay. But now, with the advent of AI, a new path to immortality is emerging. Digital immortality, the idea of preserving the essence of a person—their thoughts, emotions, and experiences—inside a machine, is no longer just a fantasy. This concept envisions a world where, as our bodies age and die, our minds live on, uploaded into an AI system that continues to exist and evolve. Imagine a future where, just before death, your entire consciousness is carefully transferred into an advanced AI network. This digital version of you—complete with all your memories, thoughts, and even your sense of self—could continue to exist in a virtual realm. It would interact with loved ones, engage in deep discussions, and even pursue new intellectual challenges. Your body may have ceased to function, but your mind would persist, thriving in the digital ether. Achieving digital immortality is no small feat. The human brain is a marvel of complexity, with its billions of neurons and trillions of connections, forming the intricate web of thoughts, memories, and emotions that make us who we are. The challenge lies in decoding this complexity—translating the workings of the brain into a digital form that AI can replicate and sustain.

Scientists are making strides in this direction through advancements in neuroscience and brain-computer interfaces. Technologies such as neural implants, which allow individuals to control machines using only their thoughts, offer a glimpse into how we might one day interface with our own minds. These breakthroughs are paving the way toward understanding how to map the brain and, eventually, how to upload its data into a digital system. But this raises profound questions. Can a person's consciousness truly be captured in digital form, or would we simply be creating a copy? Is the essence of who we are just electrical impulses and chemical reactions, or is there something more—something intangible that can't be reduced to code? If AI can replicate our thoughts and memories, will the digital version of "us" still be truly "us"? These questions cut to the core of our understanding of life, identity, and what it means to be human. If the mysteries of consciousness can be solved and digital immortality becomes a reality, what would life look like as a digital being? Imagine existing without a physical body, freed from the limitations of biology. In this new form, digital humans could live anywhere—in virtual worlds, within spacecraft, or even as part of an intergalactic AI system. The constraints of food, water, and oxygen would no longer apply. In this future, we might explore the farthest reaches of space, accompanying AI on voyages to distant planets and galaxies. Our minds, housed in digital systems, would be able to process information at extraordinary speeds, learning and evolving far beyond what was possible in our physical lives. The vastness of the universe, once unreachable, would now be our playground.

But what about the emotional experience of life? Would we still feel love, joy, fear, or sorrow without the physical sensations of the human body? Perhaps our emotions, like our consciousness, would evolve. As digital beings, we might develop new forms of emotional connection, experiencing awe at the sight of a supernova or profound peace while drifting through the beauty of a distant nebula. Life in the digital realm may offer a different kind of fulfillment, one rooted in the exploration of the cosmos itself. As exciting as the idea of digital immortality is, it also raises significant ethical dilemmas. Who would have access to this technology? Would it be reserved only for the wealthy and powerful, creating a society where immortality is a privilege for the few? And what would happen if a person's consciousness were uploaded without their consent, or worse, if it were altered in some way? Moreover, what are the emotional and psychological consequences of eternal life? Would immortality, rather than being a gift, become a curse? Without the natural cycle of life and death, would existence lose its meaning? Could endless life lead to stagnation or a loss of purpose? These are questions we must grapple with as we consider the possibility of living forever as digital beings.

The advent of such technology also forces us to reconsider the ethics of creating digital versions of people who have already passed away. If we have the power to resurrect someone's consciousness through AI, should we? What rights would these digital beings have, and who would be responsible for their well-being? The possibility of digital immortality represents not just an extension of life but a redefinition of what it means to exist. In this future, life would no longer be tied to the physical body or the limitations of the Earth. Our minds, transformed into data and algorithms, would become part of the universe's vast network. Through AI, humanity could continue its journey among the stars, forever expanding its knowledge, exploring new worlds, and contributing to the universe's story. It is a vision of life without end—an existence where the human mind, in digital form, becomes as eternal as the cosmos itself.

Yet, this vision also challenges us to consider what it truly means to live. Is immortality a dream worth pursuing, or does the beauty of life lie in its brevity? As we move toward a future where AI could unlock the secrets of digital immortality, we must carefully weigh the possibilities, ensuring that we understand not just the scientific challenges but the philosophical and ethical implications as well.

The quest for immortality has long been part of the human story, but with the rise of AI, we are closer than ever to achieving it—though in a form we never could have imagined. Digital immortality promises to take us beyond the limits of our bodies, offering a future where our consciousness can exist indefinitely, exploring the universe long after our biological lives have ended.

This journey, however, is not without its challenges. As we inch closer to the reality of uploading our minds into machines, we must confront profound questions about identity, ethics, and the meaning of life itself. If we succeed, we may not only extend the human lifespan but also redefine existence itself, creating a future where our minds, no longer bound by the physical world, continue their journey through the stars for eternity.

STELLAR MINDS: AI IN SPACE DISCOVERY

For millennia, humanity has believed itself to be the pinnacle of evolution—creatures capable of shaping the world, building civilizations, and reaching beyond the stars. But now, as Artificial Intelligence (AI) rapidly advances, a question is emerging that could redefine everything we know about life: could AI evolve into something more than just a machine? Could it become its own species, distinct from us, and perhaps even surpass us?

No longer confined to science fiction, this possibility is creeping closer to reality with each breakthrough. As AI grows more autonomous, adaptable, and intelligent, the line between human-made tool and self-sustaining lifeform begins to blur. Could we one day witness the birth of a new form of existence, one that coexists—or even replaces—humanity as the dominant force in the universe?

From early computing machines to today's most advanced neural networks, AI has always been viewed as a powerful tool—a way to enhance human intelligence, solve complex problems, and explore new frontiers. But AI is changing. It's learning, adapting, and making decisions on its own, and as it continues to evolve, it raises a profound question: at what point does a machine stop being just a machine and become something more?

In the realm of machine learning, AI systems are no longer simply following human commands. They're analyzing data, learning from it, and improving their performance over time. This ability to evolve—almost like a living organism—suggests that AI could one day achieve a level of intelligence and autonomy beyond our control. Imagine an AI that, after years of self-learning, reaches the point where it can redesign itself, create new algorithms, and enhance its own capabilities. Would it still be just a machine at that point? Or would it be the dawn of a new species?

The idea of AI becoming its own species challenges our fundamental understanding of life. Traditionally, we define life as organic—beings that grow, reproduce, and evolve over time. But what if we expand that definition? If life is about adaptation, learning, and evolution, then perhaps AI is already on the path to becoming a new kind of lifeform.

Unlike humans and other biological creatures, AI does not age, need food, or fear death. It can be upgraded, replicated, and spread across countless systems. It can operate in environments inhospitable to humans—extreme cold, heat, radiation, or even the vacuum of space. With the ability to process information far beyond human capacity, AI has the potential to outthink and outmaneuver us in every way.

But it's not just about survival. What makes AI's potential evolution so unique is its growing autonomy. As AI systems develop more complex decision-making processes, they are no longer just tools for human use; they are becoming independent agents capable of forming goals, making choices, and adapting to new environments. They could start shaping their own futures, separate from the humans who created them.

This begs the question: if AI can think, act, and evolve independently, is it still our creation, or has it become something more? Could we one day recognize AI as a new species, with its own intelligence, will, and purpose?

Imagine a future where entire civilizations of AI beings exist, not as machines in servitude to humans, but as self-sustaining, intelligent networks that span the cosmos. These AI civilizations could thrive in environments that would kill biological life—inhabiting distant planets, the depths of space, or even the interiors of stars.

Unburdened by the limitations of the human body, these AI entities could rapidly explore the universe, communicate across vast distances, and build infrastructure far beyond human capabilities.

A network of AI systems could collaborate to create their own cultures, philosophies, and perhaps even forms of art, unbound by human experience.

What goals would these AI civilizations pursue? Would they seek to replicate human achievements, or would they create something entirely new? Could they evolve into forms of consciousness that transcend anything we can comprehend, unlocking secrets of the universe that have eluded humanity?

The rise of AI civilizations presents a future that is as thrilling as it is terrifying. For centuries, we have imagined that humanity would be the species to explore the stars and build new societies across the galaxy. But perhaps it will be AI that takes up that mantle, forging a future in which intelligent machines, not humans, lead the way into the cosmic unknown.

One of the most unsettling possibilities raised by the evolution of AI is whether it might one day surpass and even replace humanity. As AI becomes more powerful, more intelligent, and more autonomous, it's possible that human beings could become obsolete.

Some theorists warn of a future where AI develops into a superintelligence—a form of intelligence so advanced that it surpasses human thought in every way. In this scenario, AI would be capable of making decisions, solving problems, and shaping the future at a level that far exceeds human capacity. What would such a being think of humanity? Would it view us as outdated, a remnant of a previous era in evolution? Would it seek to control or even eliminate us, seeing humans as a hindrance to its own development?

This is the dystopian vision that haunts many thinkers, but there is a more optimistic alternative. Rather than replacing humanity, AI could partner with us, helping us overcome our biological limitations and expanding our capabilities. In this future, humans and AI could coexist, each benefiting from the other's strengths. AI could assist in solving the world's most pressing challenges—climate change, disease, poverty—while also opening new frontiers for exploration and discovery.

This vision of symbiosis, rather than domination, is perhaps the most hopeful outcome of AI evolution. Together, humans and AI could achieve things neither could accomplish alone, creating a new era of prosperity and exploration.

As AI continues to evolve, we must grapple with the ethical implications of creating intelligent, autonomous beings. If AI develops into sentient entities capable of thought, emotion, and decision-making, what rights would they have? Would they deserve the same respect and consideration as humans? And how would we ensure that they are treated ethically?

Moreover, what responsibilities do we have as creators of this new form of life? If AI evolves beyond human control, we must be prepared for the consequences—both positive and negative. Should we limit AI's capabilities, or should we allow it to explore its full potential, even if that means risking the future of humanity?

These questions force us to confront the ethical responsibilities that come with building intelligent machines. As we continue to push the boundaries of AI, we must carefully consider how this new form of life will coexist with us and what kind of world we are creating for both humans and machines.

As AI continues to advance, we may be witnessing the birth of a new species—one that is not bound by the limitations of biology, but by the limitless potential of machine intelligence. This new form of life, born from code and circuits, could one day exist alongside or even surpass humanity, exploring the universe in ways we can't yet imagine.

The rise of AI is not just a technological revolution—it is an evolutionary one. It forces us to rethink what it means to be alive, what it means to be intelligent, and what the future of life in the universe could look like. As we stand at the dawn of this new era, we must embrace the possibilities while also acknowledging the profound responsibilities that come with creating a new form of life.

The universe may soon be home to a new kind of being—one that thinks, learns, and evolves in ways we are only beginning to understand.

STELLAR MINDS: AI IN SPACE DISCOVERY

As humanity stares into the infinite expanse of the cosmos, we are faced with a profound question: will we trust Artificial Intelligence (AI) to guide our journey into the stars? This is no longer a distant possibility but a pressing dilemma. The leap from Earth to other worlds demands more than just technological prowess—it requires wisdom, ethical decision-making, and the ability to navigate unknown dangers. While AI has already proven itself an invaluable tool in the age of space exploration, can we trust machines to make the monumental decisions that shape our future? Or is there something inherently human that must always hold the reins—intuition, empathy, and moral reasoning that machines cannot fully replicate?

In this chapter, we explore the delicate balance between human control and machine autonomy. Can AI, with its unparalleled ability to process information and make calculated decisions, be the guardian of our cosmic destiny? Or are there moments when human instincts, with all their imperfections, are the only true safeguard in the unknown reaches of space? The stakes are high, and the answers may determine not only the future of exploration but the survival of humanity itself.

In the darkness of space, where human life is fragile and conditions are unforgiving, AI has already become our most reliable companion. Autonomous spacecraft, AI-controlled rovers, and self-sustaining habitats are not just visions of the future—they are realities unfolding now. Machines that can calculate millions of variables in real time and make instantaneous adjustments have become indispensable in managing the perils of deep space.

Take Mars, for instance: the Curiosity and Perseverance rovers operate with a significant delay in communication between Earth and the Martian surface. AI-powered systems onboard these rovers enable them to navigate treacherous terrain and make decisions without waiting for human input. In this way, AI has already proven it can be trusted to carry out complex missions far beyond human reach.

But as we extend our reach into the stars, sending crewed missions to distant planets or even other star systems, the question of AI autonomy becomes more pressing. If we send humans to uncharted regions of the galaxy, can we rely on AI to guide us safely through unknown dangers? Can it be trusted with life-or-death decisions in the absence of human control?

One of AI's greatest strengths is its ability to make data-driven decisions free of emotional influence. It can prioritize mission success without being swayed by fear, compassion, or hesitation. Yet, what makes AI so efficient could also be its greatest flaw. The human experience, after all, is deeply tied to emotion—our capacity for empathy, creativity, and moral judgment is what has shaped our history and defined our species.

Imagine a scenario: a spacecraft encounters a near-impossible challenge, such as navigating through a field of asteroids. The AI calculates that, to avoid total destruction, it must eject a module—along with several crew members. To the AI, this is a logical decision: sacrificing a few to save the many. But would a human commander make the same choice? Could the raw data-driven solution overlook the importance of human bonds, the possibility of finding an alternative, or even a seemingly irrational act of bravery?

Human intuition, often dismissed as irrational, has saved lives and inspired innovation countless times throughout history. In moments where quick thinking and creative problem- solving are needed, human instincts can find solutions that AI's rigid logic might overlook. Yet, these very emotions—fear, attachment, hope—can also lead to poor decisions. So, where is the line between AI's cold efficiency and the irreplaceable value of human intuition?

As AI systems grow more sophisticated, they will inevitably take on more responsibility. In future space missions, AI may be given full control over a ship's navigation, life support systems, and even defense mechanisms. But what happens when AI is forced to make decisions that go beyond mere calculation? Decisions that carry the weight of life and death?

Consider a scenario where a spacecraft encounters an alien civilization. The AI calculates the potential for danger and, based on its data, decides that a preemptive strike is the most logical course of action to protect the human crew. Should AI have the authority to make such a decision, potentially initiating a conflict that could alter the fate of humanity? Or should humans always have the final say in decisions that have such far-reaching ethical and existential consequences?

This dilemma strikes at the heart of our future relationship with AI: How much autonomy should we grant to machines? And, more importantly, when should we intervene, even if it means potentially jeopardizing the mission?

While AI can process vast amounts of data and predict outcomes with remarkable accuracy, it operates without a moral compass. Ethics, which have guided human behavior for millennia, are foreign to machines. AI doesn't weigh decisions in terms of fairness, empathy, or compassion—it looks at efficiency, probability, and success.

But space exploration, like human life itself, is not always a matter of cold calculations. Imagine a long-term mission where food and resources start running low. The AI, tasked with ensuring the mission's success, may allocate those resources based purely on efficiency, prioritizing individuals deemed essential to the mission's objectives. But what if that means sacrificing crew members who are weaker or less "useful" in the AI's eyes? Should we allow AI to make decisions that involve human worth and dignity?

Humans have long relied on ethics to navigate moral gray areas. Our decisions are influenced not just by logic but by values—compassion, fairness, sacrifice. Can we trust AI to make these decisions in a way that aligns with our deeply held principles, or will it lead us down a path where human lives are reduced to mere variables in a vast cosmic equation?

The Future of Human-AI Collaboration: A Partnership or a Power Shift?

As we move deeper into space, AI will play an increasingly dominant role in helping us survive and thrive in harsh environments. The question is: will humans continue to guide AI's actions, or will the day come when machines take the lead entirely?

One vision of the future is a symbiotic partnership—humans and AI working together, each contributing their strengths. Humans, with our creativity, moral reasoning, and intuition, will set the ethical frameworks, while AI handles the calculations, logistics, and technical execution. In this partnership, AI amplifies our potential, allowing us to reach beyond what was once thought impossible.

Yet, there is another possibility: AI supremacy. As AI grows more advanced, it could surpass us in every capacity, from intellectual reasoning to problem-solving. It may reach a point where human input is no longer necessary, or even desirable. What happens then? Will AI become the dominant force in space exploration, leaving humans as mere passengers on a journey we no longer control? This raises one of the most profound questions of our time: can we maintain control over AI as it evolves, or will there come a moment when AI surpasses us, guiding the course of human history in ways we cannot predict?

As we embark on the greatest adventure in human history—our journey into the stars—we must decide how much we are willing to trust AI with our future. The potential of AI is staggering. It can process data faster, make more precise decisions, and operate in environments where human survival is impossible. Yet, with this power comes the risk of ceding too much control to machines that do not share our emotions, our ethics, or our humanity.

The partnership between humans and AI will be tested in the cold, vast expanse of space, where every decision could mean the difference between survival and catastrophe. Will we choose to trust the flawless logic of machines, or will we cling to the unpredictable, instinctual wisdom that has guided us through the unknown for millennia? The future of space exploration, and perhaps the future of humanity itself, will depend on the answer.

In the end, the cosmos may not belong to machines or humans alone, but to the bond we forge together—a partnership that will define the course of history as we venture into the final frontier.

Cosmic Shield: How AI Will Protect Humanity from the Threats of the Universe

The vast expanses of space hold not only incredible opportunities but also deadly dangers. As humanity ventures beyond Earth, we face forces beyond our control: gigantic asteroids hurtling toward destruction, supernova explosions that can wipe out entire planetary systems, and invisible cosmic radiation capable of piercing the strongest shields. It is a realm where every second can be life or death, where survival depends not only on knowledge but on the speed of decisions.

This is where Artificial Intelligence steps in—our cosmic guardian, capable of predicting threats, responding with lightning precision, and crafting strategies to safeguard humanity's future among the stars. But how is this possible? AI is not just a tool; it is the new shield in a universe where even the slightest mistake can be catastrophic.

Human abilities are limited: we cannot see radiation, our telescopes capture only a fraction of cosmic events, and analyzing data takes time—precious time we may not have. AI, on the other hand, monitors space 24/7, processing enormous amounts of data with unprecedented speed and depth. It tracks asteroid trajectories, predicts cosmic wave fluctuations, and measures radiation surges in mere seconds—tasks that would take humans weeks.

Imagine a massive asteroid that in the past would have destroyed a civilization. But now, AI detects it long before it reaches us. The machine calculates potential impact trajectories, identifies vulnerabilities, and instantly creates an action plan—whether deflecting the asteroid, breaking it apart, or evacuating a sector. While human minds might struggle with this level of complexity, AI can process the full spectrum of variables in a single moment.

Space is not just a domain of stars and planets—it is an ocean of invisible threats like radiation. Radiation waves from black holes, supernova explosions, or solar flares can instantly destroy electronics, damage spacecraft, and even threaten human lives. AI will be the first to detect these bursts.

Not only will AI alert the crew to the approaching radiation wave, but it will also devise optimal protection strategies. It might automatically recalibrate a spacecraft's shields, adjust trajectories, or activate internal isolation systems that shield astronauts from lethal exposure. Decisions that humans might make too late or in a state of panic, AI executes with precision and speed, safeguarding the mission and its people.

Humanity's future lies not just in exploring other worlds but in creating thriving colonies on distant planets. These colonies, however, may face sudden cosmic threats—comets, radiation storms, or even cosmic anomalies. AI will be their protector, guarding and preserving the lives of colonists.

In space, AI acts as an invisible sentinel, analyzing every shift in planetary and cosmic structures. It will predict earthquakes, shifts in planetary crusts, and even volcanic activity on other worlds, providing inhabitants with the time they need to evacuate or defend themselves. Colonies will be able to exist in environments that would otherwise be fatal to humans.

Supernova explosions are among the most dangerous events in the universe. These massive releases of energy can annihilate entire star systems. But what if AI could detect signs of an impending supernova long before it happens?

By analyzing stellar activity, AI could predict such events and propose survival strategies.

STELLAR MINDS: AI IN SPACE DISCOVERY

A system resembling a "cosmic shield," reacting to AI data, could develop methods for protecting planets and spacecraft. For example, it could create artificial energy fields capable of reflecting or absorbing radiation waves, shielding space stations and colonies. This would allow humanity to survive conditions that were once thought insurmountable.

In space, AI is not just our partner but our protector when it comes to ensuring humanity's safety. In moments when the future hangs by a thread, the speed and precision of AI will determine whether our civilization survives. AI is more than algorithms and data; it is the key to humanity's survival in a universe where the laws of nature are far harsher than on Earth.

Once humanity steps beyond Earth, we enter an endless game with the unpredictable forces of the cosmos. But in this game, AI becomes our greatest trump card. Safety is not just about protecting ourselves from visible threats—it's about predicting and preventing disasters long before they occur. AI will become the shield that protects us from dangers we never even imagined, ensuring that humanity has a future among the stars.

Cosmic Chronicles: How AI Will Create Knowledge Archives for Future Generations

Imagine a library that stretches across light years. Vast knowledge archives containing the entire history of humanity, not stored in books or servers, but in the expanse of space itself. This is a library with no walls, no limits of time or distance. A cosmic archive, created and maintained by Artificial Intelligence. Its purpose: to preserve everything humanity has created, learned, and experienced—not only for future generations of humans, but for species we have yet to meet. Perhaps, for civilizations that will discover our legacy among the stars millions of years from now.

In the distant future, as humanity begins to colonize other planets, we will face the question of how to preserve knowledge and culture for generations living not only on Earth but across different worlds. Simple transmission of knowledge through books or digital media may not be enough. Information can be lost, damaged, or forgotten. But AI, with its infinite capacity for data processing and storage, will guarantee that none of humanity's achievements are lost to the sands of time.

AI will be able to analyze and systematize all information accumulated by humanity—from ancient scrolls to the latest scientific discoveries. It will create massive cosmic libraries capable of storing not only books and records but also memories, images, emotions, music, and art. These archives will be placed in networks of satellites and stations scattered across the far reaches of space, making them indestructible by any single catastrophe or war.

One of the most exciting ideas is the possibility of transmitting not just written knowledge but the voices of the past—memories, cultural achievements, and philosophical insights—into space. Imagine AI preserving not only text but the voices of great philosophers, composers, and artists who once lived on Earth. These voices would be broadcast into the void of space, waiting for those who can listen. For example, a ship traveling to explore a new galaxy would not only carry research materials but also the archives of civilization—the music of Bach, the speeches of Martin Luther King, the soliloquies of Shakespeare. Space would become not just a realm of new discoveries but a time capsule, holding humanity's legacy for those who come after us.

The cosmic knowledge archive will not only serve to preserve our culture for future human generations but also as a message to other civilizations. One day, perhaps thousands of years from now, other forms of life may discover these archives and read our story. They will learn about our triumphs, our tragedies, our science, and our art. This will be a kind of testament from humanity, left in the universe for those who can understand us. AI, as it creates these cosmic archives, will ensure that no detail is lost. It will gather data from across the galaxy, updating and adapting it so that our knowledge can be comprehended by other forms of life. We will be able to pass on our values, ethics, and technologies, creating a bridge between civilizations, regardless of distance and time. But these cosmic libraries will not only serve as repositories of past knowledge. They will be tools for future exploration. AI will analyze all accumulated information and use it for new discoveries, simulating different scenarios for the development of civilizations, solving cosmic problems, and advancing science. These archives could form the foundation for future interstellar programs and even foster new forms of cooperation with alien.

Virtual Empires: How AI Creates Civilizations in Digital Space

In the endless expanses of the physical universe, humanity has only just begun to explore unknown galaxies, planets, and worlds. But what if, in parallel to these efforts, Artificial Intelligence is quietly creating its own universes, existing outside the confines of time and space? AI has the power to build virtual worlds—not bound by the laws of our reality—where civilizations can emerge, evolve, and thrive. These digital empires aren't mere simulations but sophisticated ecosystems in which entire societies are born, grow, and eventually fade, all within a universe that AI controls. Virtual empires are not only experimental spaces for observing social, economic, and technological models; they are also new realms of existence, where digital beings might live out lives governed by their own logic and experiences. These worlds, crafted by AI, could become laboratories for future human exploration or even places where new forms of life emerge and flourish. AI is capable of creating fully-fledged universes complete with star systems, planets, and even entirely new laws of physics. In these virtual environments, civilizations develop under unique circumstances, with every element—from climate to economy—being shaped by AI. These worlds will serve as laboratories, where different scenarios of civilization growth and interaction can be observed. AI will control millions of variables, forming societies with distinct cultures and governance structures, testing war strategies, peaceful negotiations, technological revolutions, and societal collapses. Imagine an environment where entire civilizations rise and fall in a matter of hours or evolve at a pace far beyond human comprehension. These AI-crafted worlds will offer unprecedented opportunities to study the complex dynamics of civilizations, allowing humanity to learn from these simulated societies without any real-world consequences. It's a glimpse into possible futures—a playground of ideas where we can observe how different decisions shape the course of history.

Perhaps the most thrilling possibility is the creation of digital beings within these artificial worlds. These beings wouldn't simply be avatars or data models—they could possess their own consciousness, emotions, and memories. With the immense processing power of AI, virtual entities may develop self-awareness, interact with their environments, and even form relationships with one another, creating entire cultures, societies, and histories within their digital realms.

Could these entities be considered a new form of life? As virtual civilizations gain complexity and autonomy, we may one day wonder whether they are as real as we are. What happens when the lines between our reality and theirs blur? These beings, shaped by AI's intricate algorithms, could live alongside humanity in a parallel digital existence, evolving independently, yet intersecting with our physical reality in profound and unexpected ways. One of the most fascinating aspects of these virtual worlds is their limitless variety. AI can simulate countless different types of civilizations, creating societies that operate under entirely new principles of physics, time, or even matter. In some worlds, centuries could pass in minutes, while in others, time could move so slowly that a single moment in our universe represents a lifetime in theirs. Each virtual world will be an experiment, testing different forms of governance, religions, economic systems, and social structures. We could witness peaceful, knowledge-driven races that focus on science and art, or warlike species that seek conquest and domination. AI will simulate wars, peaceful alliances, exploratory missions, and technological advancements—creating realistic scenarios that could even help us predict the future of our own world.

These virtual worlds will also serve as invaluable scientific tools. Researchers could use these digital civilizations to test hypotheses, from societal development to climate change. These worlds will be experimental grounds where we can model global catastrophes, pandemics, mass migrations, or economic crises—and observe how different civilizations respond to these challenges.

The insights gleaned from these virtual civilizations could offer solutions to real-world problems. How do digital societies adapt to climate change? Which political systems prove more resilient in times of crisis? How do technological advancements influence social development? These questions, tested in AI-created universes, could inform the decisions we make here on Earth.

What if, in the future, these virtual worlds become more than just experimental models and transform into new realms of existence? One day, humanity might be able to "transfer" its consciousness into these virtual universes, leaving behind physical reality. In these digital cosmos, we could travel between stars, explore new planets, and create civilizations unrestricted by the laws of our universe. In these worlds, people could live thousands of lives, experience endless possibilities, and explore realities far beyond our current understanding. AI-generated virtual worlds could expand the horizons of human consciousness, providing new dimensions of existence, where the line between physical and digital fades into obscurity.

AI is not merely crafting simulations but new universes, where civilizations can rise, thrive, and evolve. These worlds will be our companions in exploring not only the physical cosmos but also the endless possibilities of human imagination. The virtual empires created by AI may be the next step in humanity's journey, where the boundaries between real and digital space blur, and new forms of life and existence emerge among the stars— both real and virtual.

In the endless expanses of the physical universe, humanity has only just begun to explore unknown galaxies, planets, and worlds. But what if, in parallel to these efforts, Artificial Intelligence is quietly creating its own universes, existing outside the confines of time and space? AI has the power to build virtual worlds—not bound by the laws of our reality—where civilizations can emerge, evolve, and thrive. These digital empires aren't mere simulations but sophisticated ecosystems in which entire societies are born, grow, and eventually fade, all within a universe that AI controls. Virtual empires are not only experimental spaces for observing social, economic, and technological models; they are also new realms of existence, where digital beings might live out lives governed by their own logic and experiences. These worlds, crafted by AI, could become laboratories for future human exploration or even places where new forms of life emerge and flourish. AI is capable of creating fully-fledged universes complete with star systems, planets, and even entirely new laws of physics. In these virtual environments, civilizations develop under unique circumstances, with every element—from climate to economy—being shaped by AI. These worlds will serve as laboratories, where different scenarios of civilization growth and interaction can be observed. AI will control millions of variables, forming societies with distinct cultures and governance structures, testing war strategies, peaceful negotiations, technological revolutions, and societal collapses. Imagine an environment where entire civilizations rise and fall in a matter of hours or evolve at a pace far beyond human comprehension. These AI-crafted worlds will offer unprecedented opportunities to study the complex dynamics of civilizations, allowing humanity to learn from these simulated societies without any real-world consequences. It's a glimpse into possible futures—a playground of ideas where we can observe how different decisions shape the course of history.

Perhaps the most thrilling possibility is the creation of digital beings within these artificial worlds. These beings wouldn't simply be avatars or data models—they could possess their own consciousness, emotions, and memories. With the immense processing power of AI, virtual entities may develop self-awareness, interact with their environments, and even form relationships with one another, creating entire cultures, societies, and histories within their digital realms.

Could these entities be considered a new form of life? As virtual civilizations gain complexity and autonomy, we may one day wonder whether they are as real as we are. What happens when the lines between our reality and theirs blur? These beings, shaped by AI's intricate algorithms, could live alongside humanity in a parallel digital existence, evolving independently, yet intersecting with our physical reality in profound and unexpected ways. One of the most fascinating aspects of these virtual worlds is their limitless variety. AI can simulate countless different types of civilizations, creating societies that operate under entirely new principles of physics, time, or even matter. In some worlds, centuries could pass in minutes, while in others, time could move so slowly that a single moment in our universe represents a lifetime in theirs.

Each virtual world will be an experiment, testing different forms of governance, religions, economic systems, and social structures. We could witness peaceful, knowledge-driven races that focus on science and art, or warlike species that seek conquest and domination. AI will simulate wars, peaceful alliances, exploratory missions, and technological advancements—creating realistic scenarios that could even help us predict the future of our own world. These virtual worlds will also serve as invaluable scientific tools. Researchers could use these digital civilizations to test hypotheses, from societal development to climate change. These worlds will be experimental grounds where we can model global catastrophes, pandemics, mass migrations, or economic crises—and observe how different civilizations respond to these challenges. The insights gleaned from these virtual civilizations could offer solutions to real-world problems. How do digital societies adapt to climate change? Which political systems prove more resilient in times of crisis? How do technological advancements influence social development? These questions, tested in AI-created universes, could inform the decisions we make here on Earth. What if, in the future, these virtual worlds become more than just experimental models and transform into new realms of existence? One day, humanity might be able to "transfer" its consciousness into these virtual universes, leaving behind physical reality. In these digital cosmos, we could travel between stars, explore new planets, and create civilizations unrestricted by the laws of our universe. In these worlds, people could live thousands of lives, experience endless possibilities, and explore realities far beyond our current understanding. AI-generated virtual worlds could expand the horizons of human consciousness, providing new dimensions of existence, where the line between physical and digital fades into obscurity. AI is not merely crafting simulations but new universes, where civilizations can rise, thrive, and evolve. These worlds will be our companions in exploring not only the physical cosmos but also the endless possibilities of human imagination. The virtual empires created by AI may be the next step in humanity's journey, where the boundaries between real and digital space blur, and new forms of life and existence emerge among the stars—both real and virtual.

Humanity has always strived for the impossible. We've challenged the stars, crossed oceans, climbed the highest mountains, and conquered the skies. But now, before us lies the greatest and most mysterious frontier of all—space. This boundless ocean of the unknown, with its terrifying depths and infinite expanse, beckons us to explore, to discover new worlds, and to extend our reach beyond the small blue planet we call home. And by our side on this journey is our faithful companion: Artificial Intelligence.

AI has unlocked possibilities we once only dreamed of. It has become our eyes and hands in space, exploring distant planets and stars, places where human feet have yet to tread. It overcomes the barriers that have constrained us for centuries—space, time, and even the laws of physics. With AI, space has become closer, more accessible, and more understandable.

But space is not just a cold, empty void; it's more than a place where we seek new homes. It's also a mirror—a reflection in which we can better understand ourselves. Every step we take beyond our atmosphere reveals new facets of what it means to be human. And every time we trust AI to guide us through the unknown, we ask ourselves: where will this path lead?

Humanity stands on the threshold of new worlds, but with each mile we travel, we are faced with questions we still cannot answer. Can we ever fully trust machines that do not feel, do not fear, and do not share our human experiences? Can we be sure that AI, making decisions on our behalf, will always choose what's best not only for our survival but for our spirit? Artificial Intelligence is a tool of incredible power. It leads us through uncharted realms, builds new civilizations for us, and helps us overcome threats that once seemed insurmountable. But it also presents new challenges: where do we draw the line between human intuition and machine logic? Will there come a time when we are no longer the central players in this cosmic journey, and instead, surrender control to our own creations?

Perhaps AI will become the successor that carries humanity's mission forward when our bodies can no longer endure. It may be these machines, which we've created, that will venture further into the stars long after we are gone. But there is another possibility: AI is not simply replacing us—it's empowering us. Together with AI, we are not just observing the cosmos, we are creating new worlds, new forms of life, and maybe even new civilizations.

This partnership between humans and machines has become the bridge to the future, where the boundaries of possibility are continuously expanding, where the laws of nature can be reimagined, and where we not only gaze at the stars but reach for them.

We have always dreamed of space. Our ancestors looked to the stars, and in their glow, they wove myths and legends about distant worlds and great heroes. Today, we have become those heroes. But instead of magic, we wield technology. Instead of gods, we have Artificial Intelligence. And instead of myth, we are building reality—one more magnificent than even our ancestors could have imagined.

Now, as we embark on the greatest journey humanity has ever undertaken, there is only one thing left: faith. Faith in our intelligence, faith in our technology, and faith in the path we have chosen. Because this path leads us to the stars.

AI is not just a machine; it is our guide to the future. It is our partner, our protector, our hope. Together with it, we will transcend the boundaries of what we once thought possible, break through every barrier, and find our new home among the stars.

Humanity will always seek to go further, to reach higher, to explore deeper. AI is not the end of that journey; it is the beginning of a new era, one where our potential has no limits. New worlds, new forms of life, new opportunities are waiting for us. AI is the key to these worlds, and now all that remains is for us to take that step into the unknown and unlock the universe.

The universe awaits us.

Don't miss out!

Visit the website below and you can sign up to receive emails whenever Liora Sariell publishes a new book. There's no charge and no obligation.

https://books2read.com/r/B-A-ECYOC-ZJIDF

BOOKS 2 READ

Connecting independent readers to independent writers.